INTERNATIONAL SERIES OF MONOGRAPHS IN
NATURAL PHILOSOPHY

GENERAL EDITOR: D. TER HAAR

VOLUME 47

THE TITIUS–BODE LAW OF PLANETARY DISTANCES: ITS HISTORY AND THEORY

THE TITIUS-BODE LAW OF PLANETARY DISTANCES: ITS HISTORY AND THEORY

MICHAEL MARTIN NIETO

The Niels Bohr Institute
University of Copenhagen
Copenhagen, Denmark

and

Los Alamos Scientific Laboratory
Los Alamos
New Mexico

PERGAMON PRESS

OXFORD · NEW YORK · TORONTO
SYDNEY · BRAUNSCHWEIG

Pergamon Press Ltd., Headington Hill Hall, Oxford

Pergamon Press Inc., Maxwell House, Fairview Park, Elmsford,
New York 10523

Pergamon of Canada Ltd., 207 Queen's Quay West, Toronto 1

Pergamon Press (Aust.) Pty. Ltd., 19a Boundary Street,
Rushcutters Bay, N.S.W. 2011, Australia

Vieweg & Sohn GmbH, Burgplatz 1, Braunschweig

First edition 1972

Library of Congress Catalog Card No. 78–178682

Printed in Hungary

08 016784 5

To

Allan M. Russell and Peter A. Carruthers,

— once advisors, now friends —

and to my family

... and so there ain't nothing more to write about, and I am rotten glad of it, because if I'd a knowed what a trouble it was to make a book I wouldn't a tackled it and ain't a-going to no more.

Mark Twain (Samuel Langhorne Clemens, 1835–1910), from *The Adventures of Huckleberry Finn*, 1884.

Contents

Foreword

THE undertaking of this project on the Titius–Bode Law is due to a curiosity that was originally generated while I still was in grade school when I first read about "Bode's Law". Since it was my visit to The Niels Bohr Institute that gave me the opportunity to investigate this old and puzzling question of planetary astronomy, I should first express my thanks to the Institute for its support and also to the staff of the Institute for its help in the preparation of this book.

My thanks are also directed to the many individuals and organizations, too numerous to mention in entirety, who kindly took of their time to help me in obtaining particular pieces of biographical or bibliographical information; and also to others with whom I have had discussions on scientific aspects of this topic. With apologies to those whom lack of space forbids mentioning, I would like to specify some of them.

I first express my gratitude to the staffs of The Niels Bohr Institute library, the science library of the University of Copenhagen, and the libraries of the University of California for their help in my bibliographical searches. Thanks are due to Judy Fox who, in between bearing her first child and raising a pet boa constrictor, managed to type most of the main text and to Merrie Walker who prepared the remainder of the materials. Jean Pope of the editorial office at Pergamon Press should be mentioned for her gracious and enjoyable handling of the manuscript and my dealings with the publisher.

In addition, many people deserve my appreciation for the benefit of their advice and wisdom on scientific, historical, language, and presentation matters. Among them have been Ernest Courant, Alfred S. Goldhaber, D. ter Haar, Alex Lande, Bernhard Mühlschlegel, Stanton J. Peale, L. Rosenfeld, C. Keith Scott, and Bengt Strömgren.

In this respect, I would like to single out Bengt Gustafsson for special thanks. While sharing an office with him in Copenhagen, I took greatly of his valuable time and energy for everything from discussions on scientific points to the reading of the manuscript. Whatever merits this book may have are due in no small part to his efforts.

From whatever source, it is probably clear that errors of commission and/or omission will inevitably appear in the final product. I would appreciate their being brought to my attention and take responsibility for them. In this respect, I point out that my "free form" translations from German to English are accompanied by the original texts.

It is amusing to me that the pleasure of completing this work is partly one of relief. I imagine that Truman Capote must have had similar emotions when he returned to New York City after his long stay in Kansas working on the journalistic novel, *In Cold Blood*. At least it seems so, for he was quoted as having said, "Well, back to normal now, sex and sin."

23 *December* 1970   MICHAEL MARTIN NIETO

Department of Physics,
University of California,
Santa Barbara, California

WHILE this volume has been in press, additions have been made to allow the work to be as complete as possible. In particular, the literature has been covered up to the middle of 1971. I would like to add final thank yous to Virginia Helen Curl Thompson, Merete Henriksen, and Pat Maguire for, among other things, their respective assistances on matters of statistics, bibliography, and correcting proofs.

3 November 1971
Department of Physics,
Purdue University,
Lafayette, Indiana

M. M. N.

PLATE I.

Johann Daniel Titius von Wittenberg (1729–96). A 1770 portrait
by Benjamin Calau, engraved by S. Kalle of Berlin. *Courtesy*—(Wittenberg-Halle) Universitäts- und Landesbibliothek, Sachsen-Anhalt
in Halle (Saale).

PLATE II.

Johann Elert Bode (1747–1826). Artist unknown. *Courtesy*—Zentrales Arkiv der Deutsche Akademie der Wissenschaften zu Berlin (DDR), Bildersammlung.

CHAPTER 1

Introduction

Two centuries have now elapsed since Johann Daniel Titius von Wittenberg (1729–96) (see Plate I)[†] **(1.1–1.4)** in 1766 first inserted his famous observation on planetary distances into a German translation **(1.5)** of Charles Bonnet's (1720–93) *Contemplation de la Nature* **(1.6)**. This observation can be stated in the following way: If the radius of the Earth's orbit is normalized to 10,[‡] then the radii of all planetary orbits can be given by

$$r_n = 4 + 3 \times 2^n, \tag{1.1}$$

where n is $-\infty$ for Mercury, and $0, 1, 2, \ldots$ for succeeding planets (see Table 1.1). Ever since, this "Law" has been clouded by controversy for two reasons.

The first is the series of events that led to Johann Elert Bode (1747–1826) (see Plate II) being associated with the Law instead of its author, Titius. For most of the last century it was called Bode's Law; and even today, in both textbooks and research articles, Bode is still most often given credit. Even if it is labeled as the Titius–Bode Law, it is quite clear that most authors, both scientists and historians of science, are not acquainted with the exact details behind the dual title.

The second and more fundamental controversy is whether there is any physical significance to the Law; i.e. some dynamical reason that will explain the orbits as eigen-solutions, like quantum mechanics explains the (Niels Henrik David, 1885–1962) Bohr atomic orbits. This problem has led to much discussion during the past two cen-

---

[†] Much of the biographical data in this book is taken from refs. **(1.1–1.4)**.

[‡] Historically, a normalization to 10 was proposed. Often people use the more rational normalization to 1 astronomical unit (AU). We will use both, as convenient, but no confusion should arise.

1

TABLE 1.1

A comparison of the original Titius–Bode Law with observation.

The planets in parentheses were not known in 1766. The values predicted by Kepler's geometric construction and those noted by von Wolf (see Chapter 3) are also given.

Planet	n	Distance	Titius–Bode Law	Kepler	von Wolf
Mercury	$-\infty$	3·9	4	5·6	4
Venus	0	7·2	7	7·9	7
Earth	1	10·0	10	10·0	10
Mars	2	15·2	16	12·6	15
(Ceres)	3	(27·7)	28		
Jupiter	4	52·0	52	37·7	52
Saturn	5	95·5	100	65·4	95
(Uranus)	6	(192·0)	196		
(Neptune)	7	(300·9)	388		
(Pluto)	8	(395)	772		

turies. Bode, himself, was quite convinced that there was a physical reason; but in the Law belief has waxed and waned.

Today, feelings on the subject are quite divided. This can be seen by the widely different discussions that have been given on it. For example, the Law has been viewed as a mathematical curiosity (1.7), as a case history in the philosophy of science (1.8), and as a valid subject for an admittedly subjective statistical analysis (1.9) which itself was the subject of heated criticism (1.10, 1.11). As for astronomers, some are very skeptical and consider all such "laws" to be numerical coincidences and nothing else. However, even if it is only due to historical inertia, invariably a book or article on the solar system gives at least a passing reference to this subject. Furthermore, in recent decades much effort has been spent in trying to place the Law on a sound theoretical basis.

Until now, however, no thorough investigation has been made into the Law. This lack is what we intend to remedy. As our title suggests, first we shall look at the history behind the Law, see how it was modified, extended to satellite systems, and more generally formulated as new astronomical observations became available.

In Chapter 2 we shall present the prehistory of the Law from Kepler to Kant. During this period the belief that regularities existed among the planetary orbits became ingrained among astronomers. This was important, for it set up the intellectual milieu from which the Titius Law was to come. In Chapter 3 we will discuss in some detail the circumstances under which Titius presented his Law to the public, how Bode read it and included it in his own writings, and how Bode eventually became identified as the Law's author.

The next two chapters describe the Law's early successes and failures. When the orbits of Uranus and the asteroid belt were found to fall into place, the Law was highly accepted. But then questions arose concerning the lack of planets between Mercury and Venus, and later the lack of agreement between the Law and the orbit of Neptune, let alone Pluto.

We then discuss in Chapter 6 a series of observations by a number of authors over the course of a century. These dealt with modifications and generalizations of the Law to include the orbits of the satellite systems of the major planets and even comets. What is interesting is that these authors were ignorant of most of the work of their predecessors.

In Chapter 7 we come to two essentially equivalent formulations of the Law done in this century by Miss Mary Adela Blagg and D. E. Richardson. These formulations include the satellite systems of Jupiter, Saturn, and Uranus within their framework and are each a geometric progression in 1·73 (*not* 2) multiplied by a periodic function of the planet number. They are the best phenomenological representations of distances with which to investigate the theoretical significance of Titius–Bode type Laws.

In the next three chapters we discuss the significance of the Law with respect to the origin of the solar system. We make the proposals that the solar system had a nebular period, that the geometric progression originated in it, and that the progression was caused by some fluid and/or magnetohydrodynamical process. Further, we propose

3

that the periodic or "evolution" function represents a tendency towards commensurability due to a point gravitational or tidal evolution. (However, we will keep in mind the other main possibility, that the entire Law is due to some gravitational or tidal evolution process.)

The following four chapters discuss the theories of all types that have been proposed to explain the "classical Titius–Bode Law", i.e. the geometric progression. For purposes of discussion these theories are divided into three somewhat-overlapping categories: electromagnetic, gravitational, and nebular theories.

Our final chapter reviews the conclusions we have reached and emphasizes the places where future work should be focused.

As one can see, Chapters 2–6 concentrate a great deal on the historical aspects of the Law, and so the reader should have no trouble following the discussion. The next four chapters become more technical as we look at recent discussions of the Law in the light of modern astrophysical knowledge of the solar system. However, the interested reader should generally be able to follow these chapters too, for much of the discussion is on physical grounds.

It is only in Chapters 12–14 that the discussion becomes quite technical. As mentioned, it is here that we investigate in some detail the theories that have been proposed to explain the geometric progression of the Titius–Bode Law. So, as an aid to those readers who are not interested in following all of the details, in Chapter 11 we have included a summary of the contents of Chapters 12–14. Further, in Chapters 12–14 we have tried to clarify the discussion of those relevant parts of the theories that were confusing in the original papers. In any event, an essentially complete set of references has been included.

While mentioning the bibliography, it should be pointed out that in spite of the approximately 300 different letters, articles, books, and bibliographies on specialized topics that are mentioned, the particular reader may want to delve deeper into the literature in an area that could not be discussed in detail, simply because this book touches so many topics. To do this, in addition to searching through standard bibliographical sources such as *Physics Abstracts*, *Astronomischer Jahresbericht*, and *Astronomy and Astrophysics Abstracts*, I highly recommend consulting *Bibliographie Générale de l'Astronomie jusqu'en 1880* **(1.12)** and the *Royal Society Catalogue of Scientific Papers 1800–1900* **(1.13)** for works before 1900. Further, the recently completed *Astronomy and Astrophysics, A Bibilographical Guide* **(1.14)** is an important and major new contribution to aid one in "getting into" all of the literature, from antiquity up to 1970. Finally, the *Guide to Reprints* **(1.15)** is quickly becoming a valuable tool, since each year further older works and journals are being reprinted.

CHAPTER 2

Historical Background

THE foundation of the Titius–Bode Law goes back to the appearance in 1595 of Johann Kepler's (1571–1630) first famous work, *Mysterium Cosmographicum* (2.1). It was here that Kepler's unique mind was first widely exhibited in detail. With fascination one can observe his determination to get exact agreement between "theory" and precise astronomical measurements, but yet also his struggle to find a metaphysical theory as the ultimate explanation of the physical universe[†] (2.2).

As is well known, Kepler first attempted to explain the relative sizes of the planetary orbits in terms of celestial spheres and regular polygons[‡] (2.3). He envisioned a sphere for the orbit of Saturn which was inscribed by a cube which in turn was inscribed by the orbit sphere of Jupiter. The rest of the planetary orbits were described by successive spheres with an interspersed dodecahedron, icosahedron, and octahedron respectively. The agreement with observation was not especially good (see Table 1.1), but it started Kepler on his long quest to understand the planetary orbits.

When, just before death, Tycho Brahe (1546–1601) ceded his many years' observations to Kepler, Kepler was given the tool he needed to succeed. Brahe was history's greatest naked-eye astronomer and his accurate observations were the key which allowed Kepler to understand that ellipses were needed to describe orbits. Slowly Kepler discovered what we now call his First Two Laws of Planetary Motion (2.4) and his work culminated with the appearance of the Third Law in *Harmonices Mundi* (2.5), his unique union of music and astronomy.

Although some years were needed for the total significance of

[†] A complete bibliography of the publications by, of, and on Kepler can be found in ref. (2.2).

[‡] A lively account of this can be found in ref. (2.3).

Kepler's work to be realized, it ultimately produced a profound effect. For our purposes the importance was two-fold. First, his use of celestial spheres and regular polygons induced astronomers to think in terms of regular relationships existing between succeeding planetary orbits. Since his Laws were so exact, this lent credibility to at least the physical notions contained in his other ideas.

The second important thing was that Kepler noted that there seemed to be gaps between the orbits of Mercury and Venus and between Mars and Jupiter where two small unseen planets could exist. But he felt this would imply an indefinite number of orbits extending inwards from Mercury and out from Saturn. This picture did not appeal to him and he was convinced to reject it by the argument of Rheticus (Georg Joachim von Lauchen, 1514–76) **(2.6)** that the sacred number of planets was six. [The significance of this point has been discussed by Koyré **(2.7)**.] Luckily, though, Kepler placed the whole above argument in the introduction to *Mysterium Cosmographicum*, which was a very prominent place in his writings to put anything. In addition, after the discovery of his last Law of Planetary Motion (where the regularities and large distance between Mars and Jupiter could be seen in the "musical notes" of *Harmonices Mundi*), Kepler published a second edition of *Mysterium Cosmographicum* **(2.8)**. Thus, the idea that there existed missing planets had been firmly planted.

By the eighteenth century, Sir Isaac Newton (1642–1726) had dynamically explained not only planetary orbits but also the tides and gravity, a feat which Kepler had been the first to attempt **(2.4, 2.9)**. Thus, the Kepler Laws had been understood theoretically as well as vindicated experimentally so that Kepler's other ideas about relationships among the planetary orbits were still in vogue.

This can be seen, for example, in the writings of Christian Freiherr von Wolf (1679–1754). von Wolf felt that the orbits did exhibit some kind of regularity **(2.10)**. His most important statement, which foreshadowed the Titius–Bode Law but was not in the form of a geometric progression, was that (see Table 1.1) **(2.11, 2.12)**: "If one gives the distance from the Earth to the Sun in 10 units, then the distance to Mercury from the Sun is 4, that of Venus 7, that of Mars 15, that of

Jupiter 52, and that of Saturn 95."[†] He went on to discuss how there had to be such distances so that the planets would not be too close to each other.

Another influence of von Wolf was to pass on his cosmological ideas to his brilliant young disciple, Immanuel Kant (1724–1804).

In 1755 Kant published his theory of the universe, *Allgemeine Naturgeschichte und Theorie des Himmels* **(2.13)**. This book contained two sections, beginning on pages 18 and 163, where Kant made observations that concern us. Kant first noted that if one ignores Mercury and Mars because of their small masses, succeedingly distant planets from the Sun have increasing eccentricities. This, he felt, might indicate the existence of trans-Saturn planets which would have orbit eccentricities approaching that of the comets. Furthermore, he emphasized that the great distance between Mars and Jupiter indicated that there might be another planet between them.

This point was noticed elsewhere, too. In 1761 Johann Heinrich Lambert (1728–77) asked the following question **(2.14)**: "And who knows if there are not planets missing in the large distance between Mars and Jupiter that will be discovered?"[‡] This question was to prove to be important and the final fuels had been added to the intellectual fire, setting the stage for the quiet appearance of Titius' Law a few years later.

[†] The pertinent passage reads: "Wenn man die Weite der Erde von der Sonne in 10 Theile eintheilet, so bekommet davon die Weite des Mercurious von ihr 4, der Venus 7, des Mars 15, des Jupiters 52, des Saturnus 95." See refs. **(2.11, 2.12)**.

[‡] "Und wer weiss es, ob nicht schon Planeten mangeln, die aus dem weiten Raume, der zwischen dem Mars und dem Jupiter ist, hinweg gekommen sind."

Ces Globes, qui paroiffent *errer* dans l'armée des Cieux, font les *Planètes* dont les *principales* ont le Soleil pour centre commun de leurs révolutions périodiques, & dont les autres, qu'on nomme *fécondaires*, tournent autour d'une Planète principale, qu'elles accompagnent, comme des *Satellites*, dans la révolution annuelle.

Vénus & la *Terre* ont chacune leur Satellite. Un jour, fans doute, l'on en découvrira à *Mars*. *Jupiter* en a quatre, *Saturne*, cinq, & un *Anneau* ou Atmofphère lumineufe qui femble faire la fonction d'un amas de petites *Lunes* : placé à près de trois cent millions de lieuës du Soleil, il en auroit reçu une lumière trop foible, fi fes Satellites & fon Anneau ne l'euffent augmentée en la réfléchiffant.

Nous connoiffons dix-fept Planètes qui entrent dans la compofition de notre Syftème folaire; mais, nous ne fommes pas affurés qu'il n'y en ait pas davantage. Leur nombre s'eft fort accru par l'invention des Télefcopes : des Inftrumens plus parfaits, des Obfervateurs plus affidus ou plus heureux, l'accroîtront peut-être encore. Ce Satellite de Vénus, entrevu dans le dernier fiécle, & revu depuis peu, préfage à l'Aftronomie de nouvelles conquêtes.

Non feulement il étoit réfervé à l'Aftronomie moderne d'enrichir notre Ciel de nouvelles Planètes, il lui étoit encore donné de reculer les bornes de notre Tourbillon. Les *Comètes*, que leurs apparences trompeufes, leur queuë, leur chevelure, leur direction quelquefois oppofée à celle des Planètes, & très-fouvent différente, leurs apparitions & leurs difparitions, faifoient regarder comme des Météores

A 4

nämlich die Nebenplaneten jährlich um ihren Hauptplaneten, wie Trabanten, herumlaufen.

Venus und die Erde haben jegliche ihren Trabanten. Mit der Zeit wird man ohne Zweifel auch einen um den Mars entdecken. Jupiter hat ihrer viere, Saturn fünfe, nebst einem Ringe, oder einer leuchtenden Atmosphäre, welche die Stelle vieler kleinen Monden zu vertreten scheint. Da er beynahe dreyhundert Millionen Meilen von der Sonne entfernt ist, so würde er ein zu schwaches Licht bekommen, wenn nicht seine Trabanten und sein Ring dasselbe zurückwürfen und vermehrten.

Wir kennen siebzehn Planeten, die unser Sonnensystem ausmachen helfen; aber wir sind nicht versichert, daß ihrer nicht noch mehrere vorhanden sind. Ihre Anzahl ist seit Erfindung der Fernröhre sehr gewachsen; vielleicht wird sie noch mehr wachsen, wenn wir noch vollkommenere Werkzeuge, noch fleißigere und glücklichere Bemerker bekommen. Der Trabante der Venus, der im vorigen Jahrhunderte nur auf einen Augenblick gesehen, seit kurzem aber aufs neue erblicket worden, verkündiget der Sternkunde noch manche neue Entdeckungen.

Gebet einmal auf die Weiten der Planeten von einander Achtung; und nehmet wahr, daß sie fast alle in der Proportion von einander entfernt sind, wie ihre körperliche Größen zunehmen. Gebet der Distanz von der Sonne bis zum Saturn 100 Theile, so ist Mercurius 4 solcher Theile von der Sonne entfernt: Venus $4 + 3 = 7$ derselben; die Erde $4 + 6 = 10$; Mars $4 + 12 = 16$. Aber sehet, vom Mars bis zum Jupiter kömmt eine Abweichung von dieser so genauen Progression vor. Vom Mars folgt ein Raum von $4 + 24 = 28$ solcher Theile, darinn weder ein Haupt- noch ein Nebenplanete zur Zeit gesehen wird. Und der Bauherr sollte diesen Raum ledig gelassen haben?

A 4

Nimmer-

Nimmermehr! lasset uns zuversichtlich setzen, daß dieser Raum sonder Zweifel den bisher noch unentdeckten Trabanten des Mars zugehöre; laßt uns hinzuthun, daß vielleicht auch Jupiter noch etliche um sich habe, die bis itzt noch mit keinem Glase gesehen werden. Von diesem, uns unbekannten Raume erhebt sich Jupiters Wirkungskreis in $4 + 48 = 52$; und Saturnus seiner, in $4 + 96 = 100$ solcher Theile. Welches bewundernswürdige Verhältniß!

Es war der heutigen Sternkunde vorbehalten, nicht nur unsern Himmel mit neuen Planeten zu bereichern, sondern auch die Gränzen unsers Sonnenwirbels viel weiter hinauszusetzen. Die Kometen, welche, ihres betrüglichen Anblickes halber, ihres Schweifes, ihres haarigten Kernes, ihrer den Planeten oft entgegen gesetzten und von ihnen verschiedenen Richtung, ihres Erscheinens und Verschwindens wegen, für Erscheinungen gehalten wurden, die eine erzürnte Macht in der Luft angezündet hatte; diese Kometen sind zu planetischen Körpern geworden, deren lange Laufbahnen unsre Sternkundige berechnen, ihre entfernte Rückkehren vorhersagen, und ihren Ort, ihre Annäherungen und Entfernungen bestimmen. Vierzig dieser Körper erkennen anitzt schon die Herrschaft unsrer Sonne, und die Bahnen, welche einige von ihnen um dieselbige beschreiben, sind so sehr ausgedehnt, daß sie solche, erst nach einer langen Reihe von Jahren, oder wohl gar in vielen Jahrhunderten, einmal durchlaufen.

Gleichergestalt war es ein Vorrecht der neuern Sternkenntniß, zu zeigen, daß diese Kometen vermuthlich diejenigen Wandelsterne sind, wodurch die unzähligen Systeme so vieler Sonnen zusammen hängen, und die das eigentliche Verbindungsglied in der gesammten Kette der Sterngebäude abgeben. Denn wozu wäre der große Raum nöthig, der vom Saturn bis zum nächsten Fixsterne banten und sein Ring dasselbe zurückwürfen und vermehrten.

Wir kennen siebzehn Planeten, die unser Sonnensystem ausmachen helfen; aber wir sind nicht versichert, daß ihrer nicht noch mehrere vorhanden sind. Ihre Anzahl ist seit Erfindung der Fernröhre sehr gewachsen; vielleicht wird sie noch mehr wachsen, wenn wir noch vollkommenere Werkzeuge, noch fleißigere und glücklichere Bemerker bekommen. Der Trabante der Venus, *) der im vorigen Jahrhunderte nur auf einen Augenblick gesehen, seit kurzem aber aufs neue erblicket worden, verkündiget der Sternkunde noch manche neue Entdeckungen. **) Es war der heutigen Sternkunde vorbehalten, nicht nur unsern Himmel mit neuen Planeten zu bereichern,

A 4

*) Diese Erblickung ist noch zur Zeit nicht ausgemachet. Die neuesten Observationen des Durchganges der Venus unter der Sonnenscheibe, haben davon in allen Welttheilen nichts entdecket. So wahrscheinlich indessen die Sache ist, so hat sie sich gleichwohl noch nicht durch richtige Wahrnehmungen bestätiget. C

**) Gebet einmal auf die Weiten der Planeten von einander Achtung; und nehmet wahr, daß sie fast alle in der Proportion von einander entfernt sind, wie ihre körperlichen Größen zunehmen. Gebet der Distanz von der Sonne bis zum Saturn 100 Theile, so ist Mercurius 4 solcher Theile von der Sonne entfernt: Venus 4 + 3 = 7 derselben; die Erde 4 + 6 = 10; Mars 4 + 12 = 16. Aber sehet, vom Mars bis zum Jupiter kömmt eine Abweichung von dieser so genauen Progression vor. Vom Mars folgt ein Raum von 4 + 24 = 28 solcher Theile, darinn weder ein Haupt- noch ein Nebenplanete zur Zeit gesehen wird. Und der Bauherr sollte diesen Raum ledig gelassen haben? Nimmermehr! lasset uns zuversichtlich setzen, daß dieser Raum sonder Zweifel den bisher noch unentdeckten Trabanten des Mars zugehöre; laßt uns hinzuthun, daß vielleicht auch Jupiter noch etliche um sich habe, die bis itzt noch mit keinem Glase gesehen werden. Von diesem, uns unbekannten Raume erhebt sich Jupiters Wirkungskreis

---

PLATE V.

Pages 7–8 of the 1772 second edition of Titius' translation, showing his Law now printed as a footnote with the translator's mark. *Courtesy— Københavns Universitetsbiblioteket.*

chern, sondern auch die Gränzen unsers Sonnenwirbels
viel weiter hinauszusetzen. Die Kometen, welche, ihres betrüglichen Anblickes halber, ihres Schweifes, ihres
haarigten Kernes, ihrer den Planeten oft entgegen gesetzten und von ihnen verschiedenen Richtung, ihres Erscheinens und Verschwindens wegen, für Erscheinungen
gehalten wurden, die eine erzürnte Macht in der Luft
angezündet hatte; diese Kometen sind zu planetischen Körpern geworden, deren lange Laufbahnen unsre Sternkundige berechnen, ihre entfernte Rückkehren vorhersagen,
und ihren Ort, ihre Annäherungen und Entfernungen
bestimmen. Vierzig dieser Körper erkennen anitzt schon
die Herrschaft unsrer Sonne, und die Bahnen, welche
einige von ihnen um dieselbige beschreiben, sind so sehr
ausgedehnt, daß sie solche, erst nach einer langen Reihe
von Jahren, oder wohl gar in vielen Jahrhunderten,
einmal durchlaufen. *)

Endlich

kreis in 4 + 48 = 52; und Saturnus seiner, in
4 + 96 = 100 solcher Theile. Welches bewundernswürdige Verhältniß! T.

*) Gleichergestalt war es ein Vorrecht der neuern Stern-
kenntniß, zu zeigen, daß diese Kometen vermuthlich diejenigen Wandelsterne sind, wodurch die unzählichen Systeme
so vieler Sonnen zusammen hängen, und die das eigentliche Verbindungsglied in der gesammten Kette der Sterngebäude abgeben. Denn wozu wäre der große Raum nöthig, der vom Saturn bis zum nächsten Fixsterne vorhanden ist. Nennet der Abstand der Erde von der Sonne,
oder die Erdferne, 1; so ist der Saturn fast zehnmal weiter von der Sonne weg. Und was denket ihr, wie weit,
nach diesem Maaßstabe, der nächste Fixstern von unsrer
Sonne abstehe? Ihr erschrecket, wenn euch Bradley 400000
Erdfernen angiebt! Und ihr erschrecket annoch, wenn euch
Newton und Huygens sagen, daß dieser Abstand nur 34,
wenigstens 27 mal größer, als der Erde ihrer von der
Sonne sey. Aber wohlan! setzet diese ungeheure Weite
mit mir noch um ein gutes herunter. Nehmet an, daß der
näch-

da seyn, wo wir den Saturn sehen? Oder können nicht noch mehrere große Planeten-Kugeln jenseits der Saturns-bahn ihre weiten Kreise um die Sonne beschreiben, welche der Mensch nie sehen wird? Kann nicht noch innerhalb der Bahn des Merkurs ein Planet seinen Lauf vollführen, welcher, von dem Glanz der nahen Sonne bedeckt, der Erde nie zu Gesicht kömmt? und wozu der große Raum, welcher sich zwischen dem Mars und Jupiter befindet, wo bis jetzund noch kein Hauptplanet gesehen wird? Ist es nicht höchstwahrscheinlich, daß zum wenigsten ein Planet daselbst in der Bahn einhergeht, welche ihm der Finger der Allmacht vorgezeichnet hat? Dieser Planet kann noch größer als die Erdkugel seyn, und dennoch mit den jetzigen besten Fernröhren auf der Erde nicht erreicht werden, da uns selbst der 1200 mal größere Jupiter nur als ein glänzender Punct erscheint *.

Was

---

* Dies letztere scheint insbesondere aus dem ganz bewunderns-würdigen Verhältniß zu folgen, welches die bekannten sechs Hauptplaneten in ihrer Entfernung von der Sonne beobachten. Man nenne den Abstand des Saturns von der Sonne 100, so ist der Merkurius 4 solcher Theile von der Sonne entfernt. Die Venus 4 und 3=7. Die Erde 4 und 6=10. Der Mars 4 und 12=16. Nun aber kommt eine Lücke von dieser so ordentlichen Progreßion. Vom Mars an folgt ein Raum von 4 und 24=28 Theilen, worin bis jetzund noch kein Planet gesehen wird. Kann man glauben, daß der Urheber der Welt diesen Raum leer gelassen hat? Gewiß nicht. Von hier kommen wir zu der Entfer-

---

PLATE VI.

Pages 462–3 of the 1772 second edition of Bode's *Anleitung zur Kenntniss des gestirnten Himmels*, where, following Titius' discussion but not giving credit to him, Bode published the Law for the first time as a footnote.
Courtesy—Universitäts Bibliothek, Kiel.

Was erhält aber diese ungeheure Lasten der Weltkörper freyschwebend im Weltraum? Welche geheime Kraft belebt sie, und beflügelt ihren Lauf, daß sie in ihren Bahnen ungestört in richtigen und bestimmten Kreisen sich um die Sonne wälzen? Warum stehen sie nie ermattet stille, oder werden aus der Herrschaft der Sonne weggeschleudert; sondern fangen ihren Umlauf immer wieder an, wenn er vollendet ist? Ist es vielleicht der Wille des Schöpfers und eine unmittelbare Wirkung seiner Macht? Auf solche Art würden freylich diese schweren Fragen sehr leicht beantwortet seyn. Aber alsdenn müßte der Schöpfer jeden Augenblick Wunder verrichten, welches nicht glaublich ist. Sondern der Bauherr der Welt hat gewisse ewige und unwandelbare Gesetze in die Natur aller Körper gelegt, nach deren Vorschriften auch jene großen Planetenkugeln ihre weiten Reisen zurücklegen, ohne jemals in Unordnung zu kommen. Die Schwere nemlich ist die allgemeine Triebfeder ihrer Bewegung. Diese durchdringt alle Körper in ihren kleinsten Theilen. Vermittelst dieser Kraft wenden die Körper ein Bemühen an, sich beständig einander zu nähern, und dieses nach gewissen Verhältnissen ihrer Maßen und ihres Abstandes. So haben die Kugeln im Planetensystem eine Schwere oder Senkungskraft gegen den Mittelpunct desselben,

die

Entfernung des Jupiters durch 4 und 48=52, und endlich des Saturns durch 4 und 96=100 Theile.

CHAPTER 3

Formulation of the Law

THE fascinating story of the creation of the Titius Law began in Amsterdam in 1764 when the famous natural philosopher, Charles Bonnet, published his *Contemplation de la Nature* **(1.6)** (see Plate III). Due both to the merits of the book and the fame of the author, the work was well received. In fact, by the time a later French edition was prepared for a publication of Bonnet's collected works[†] **(3.1)**, the work had been translated into no less than four languages: German, Italian, Dutch, and English. It is, of course, the German translation which is critical to our story, for this is a translation by Johann Daniel Titius von (of) Wittenberg **(1.5)**.

In his edition, Titius was not satisfied with a simple translation. Titius decided to add notes, but he did it in what would be for us a very unusual and unassuming way. He simply inserted them in the main text. Why Titius did this is something of a mystery. Perhaps he was following the conservative tradition of knowledge being common property. I could also suggest that Titius was a shy man (he did not use great titles) and that he had the great respect for academic seniority which was and still is a hallmark of German universities. This might have made him hesitant to presume to add to the writings of the famous Bonnet. Whatever the reason, it had significant consequences.

Titius' historic insertion was in Part I, Chapter 4, between paragraphs six and eight of the translation. There **(1.5)** he inserted a new paragraph that began "Gebet einmal auf die Weiten der Planeten von ...". This insertion contained what became known as the Titius–Bode Law, and is shown in Plate IV. It is translated as follows.

[†] In the preface to this edition Bonnet observed that Titius had inserted notes into the German translation that we discuss below.

For once pay attention to the widths of the planets from each other and notice that they are distant from each other almost in a proportion as their bodily heights increase. Given the distance from the Sun to Saturn as 100 units, then Mercury is distant 4 such units from the Sun; Venus $4+3 = 7$ of the same; the Earth $4+6 = 10$; Mars $4+12 = 16$. But see, from Mars to Jupiter there comes forth a departure from this so exact progression. From Mars follows a place of $4+24 = 28$ such units, where at present neither a chief nor a neighboring planet is to be seen. And shall the Builder have left this place empty? Never! Let us confidently wager that, without doubt, this place belongs to the as yet still undiscovered satellites of Mars; let us add that perhaps Jupiter also has several around itself that until now have not been seen with any glass. Above this, to us unrevealed, position arises Jupiter's domain of $4+48 = 52$; and Saturn's at $4+96 = 100$ units. What a *praiseworthy relation* (bewundernswürdige Verhältniss)!

As can be seen in Plate III, this paragraph is not to be found in the original French edition **(1.6)** where paragraphs seven and eight begin "Nous connoissons dix-sept Planètes ..." and "Non seulement il étoit réservé à l'Astronomie moderne ...." (Titius combined paragraphs three and four when translating.)

It is interesting to note that although Titius called for a new planet in the position between Mars and Jupiter, he could not get himself to call it a "chief" planet, thinking it better to predict satellites. After all, a new "chief" planet had not been discovered in known history whereas satellites had been discovered. The fact that a satellite would never have had a stable circular orbit at that distance from Mars does not seem to have been known by or to have bothered Titius.

At first, no notice was given to the Titius Law. But, in the meantime, English and Italian editions of Bonnet's book appeared. Titius had access to the Italian translation by Abbot Lazarro Spallanzani (1729–99) where numerous notes and a preface were inserted. Emboldened by this, Titius clearly placed his old and new insertions as translator's footnotes when six years later he issued a second edition **(3.2)** of his translation (see Plate V).

The best evidence concerning what was Titius' inspiration in formulating the Law can be found in the same footnote, but in the fourth edition of the translation **(3.3)**. There Titius added a sentence saying: "This relationship and the related considerations which Herr Bonnet thought had first been observed by Herr Lambert had already been recited by Freyherr von Wolf in his German Physics more than forty years earlier."[†]

This quote shows us that, at least in 1783, Titius was aware of the work of von Wolf and Lambert and that he may have had communication with Bonnet concerning Titius' text insertions (a further cause for changing them to footnotes?). In any event, by looking at Titius' Law (Plates IV and V), we can observe that it consists first in quantifying in a geometrical, algebraic form the rough observations of von Wolf and then adding Lambert's belief in an empty orbit between Mars and Jupiter. This serves to confirm Titius' sources as those we discussed in the last chapter **(2.11, 2.14)**.

The identification of the von Wolf source was first made by Gerhard Ulrich Anton Vieth (1763–1836). Vieth **(3.4)** wanted to clear up the origin of the Law, which had been ascribed first to Bode and then to Titius in successive June and July 1801 issues of *Monatliche Correspondenz*. He had read the 1783 Titius translation **(3.3)** and then had come across von Wolf's observations in the 1741 fourth edition. (Because Vieth quoted the fourth edition, later writers were eventually led to state that von Wolf had noted the existence of a progression as far back as 1741 instead of 1723.)[‡]

Vieth considered it a shame that von Wolf had come as close as he had to formulating the Law, but yet had missed it. "What a shame that

---

[†] "Dieses Verhältniss, und dergleichen Betrachtung, welche Herr Bonnet glaubet zuerst von Herrn Lambert bemerket zu seyn, hat schon der Freyherr von Wolf vor mehr als vierzig Jahren in seiner deutschen Physik vorgetragen."

[‡] See, for example, p. 27 of ref. **(3.5)**.

one comes so near (and yet misses) the roots of many ideas."[†] But then, the history of science is full of these near-misses.

Soon after, Johann Friedrich Benzenberg (1777–1846) published an article (3.6) in which, among other things, he discussed the history of the Law from Titius up to 1803. In this paper Benzenberg gave vent to the lament that the Titius–Bode Law had an entirely German history unknown in the rest of Europe, not because it was untrue, but because it was written in German, an incomprehensible language. But Benzenberg also pointed out that the von Wolf source of Titius actually went back to 1723 and that the *Cosmologische Briefe* of Lambert (2.14) gave reference to a planet between Mars and Jupiter.

To return to 1772, Titius had just published the second edition of his translation as Johann Elert Bode was finishing the second edition of his astronomy book, (*Deutliche*) *Anleitung zur Kenntniss des gestirnten Himmels* (3.7). As chance had it, Bode came across Titius' note and was deeply struck by the agreement between this Law and the orbital radii of the then known six planets. Bode believed the Law at once and inserted it as a footnote into his text in time for publication (3.7) (see Plate VI). A translation of Bode's footnote follows:

This last appears to follow especially from the entirely *praise-worthy relation* [bewundernswürdigen Verhältniss] which the known six chief planets follow in their distances from the Sun. One calls the distance to Saturn 100, then Mercury is distant 4 such units. Venus is 4 and 3 = 7. The Earth 4 and 6 = 10. Mars 4 and 12 = 16. But now comes a gap in this so orderly progression. From Mars out follows a position of 4 and 24 = 28 units where up to now no planet is seen. Can one believe that the Creator of the Universe has left this position empty? Certainly not. From here we come to the distance of Jupiter through 4 and 48 = 52 and finally Saturn's through 4 and 96 = 100 units.

[†] "Schade! dass man so um die Genealogie mancher Idee kommt." Schade auch, dass man so um die Genealogie mancher Idee beim Übersetzen kommt.

First notice that Bode did *not* give credit to Titius. Despite this, it is clear, not only historically but even from the contents of the two formulations, that Bode took the Law from Titius. Comparing Plates IV–VI we see that Bode used the same reasoning in the same order as Titius, and further, he even described the Law by the same term, "bewundernswürdige Verhältniss". But there is *no* acknowledgement of Titius. This is contrary to common belief.

Most authors claim that Bode gave Titius credit because in later editions of this book **(3.8)**, as well as in other writings, he did. The common ninth edition **(3.8)** states quite clearly that Bode first published his formulation in the 1772 Hamburg second edition of the *Anleitung* after he had first seen it in the 1772 second edition of Titius' translation.[†] Other commentators must have seen or heard of these later sources and assumed that the acknowledgement went back to the second edition of 1772. This primary source, the second edition published by Bode himself, was overlooked because it is rare and difficult to obtain. In this manner the mistake about Bode giving Titius credit was perpetuated.

This chain of events suggests that again there might have been communication, this time between Titius and the then young astronomer Bode over the "stealing" of Titius' Law. This would explain why Bode did an about face and thereafter very conscientiously gave Titius credit. Certainly this aspect of the Law's history concerning the possibility of communications among Bonnet, Titius, and Bode is also worthy of further investigation.

There was, however, one main distinction between the two versions of the Law. Bode clearly predicted that an undiscovered "main" planet, not possibly a "neighboring" planet, was located between Mars and Jupiter. This was emphasized in the 1777 third edition of

---

[†] "Von jener Progression rede ich bereits in der zweiten Auflage dieser *Anleitung zur Kenntniss des gestirnten Himmels,* die im Jahre 1772 noch in Hamburg erschien, nachdem ich solches zuerst in der vom Prof. Titius in Wittenberg veranstalteten Uebersetzung von Bonnets *Betrachtung über die Natur.* 8. Leipzig 1772. Seite 7, gefunden und nachher auch in allen folgenden Auflagen dieser *Anleitung* vorgetragen habe."

13

the *Anleitung* **(3.9)** when Bode added the following section to his footnote:[†]

> From the Law discovered by Kepler, that the ratio of the periods of revolution of two planets behaves like the ratio of the cubes of their distances from the Sun, it can be calculated that this main planet between Mars and Jupiter must complete its revolution around the Sun in $4\frac{1}{2}$ years.

As a side note, it is amusing to speculate on what would be the modern reaction to the Law if it were presented now, but in its original context. One question would be about the deviations of the then known planets from the Law by the observational few percent (see Table 1.1). This is slightly ironical since these deviations are nothing compared to what today's theorists in high energy and astrophysics are all too often willing to call agreement with theory. Another question would be about the belief that there had to be another planet between Mars and Jupiter because the Creator of the universe would not allow an orbit to be empty. But is this not like modern science declaring, "If an equation has a solution, nature will fulfil. it" [as in (Dmitri Ivanovich, 1834–1907) Mendeleev's elements in the periodic table or Dirac's negative-energy state positrons]? Basically, the only difference is that "God" has become "nature".

All this being said, there remains the question of why Bode received the credit and not Titius. There are three main reasons, the first being the method of publication. Because of the way Titius published his Law, it was not widely brought to the attention of the astronomers of the day. Even those who might have read it certainly must have had a psychological barrier against it because of its location and its being a translator's note. Bode, on the other hand, placed it in his basic book on astronomy, as well as in other astronomical writings, which were much more "respectable" locations.

As we shall see in the next chapter, a second reason is that Bode

---

[†] "Nach einem vom Kepler erfundenen Gesetz, dass sich nemlich die Quadrate der Umlaufszeiten zweier Planeten gegen einander verhalten, wie die Würfel ihrer Entfernungen von der Sonne lässt sich berechnen, dass dieser Hauptplanet zwischen Mars und Jupiter seinen Umlauf um die Sonne in $4\frac{1}{2}$ Jahre vollenden müsste."

became something of a crusader for the Law. Although for many years, the Law was not highly thought of, Bode mentioned it time and again in his many writings, whereas Titius was not heard from. Thus, when the asteroids and Uranus were discovered, the Law became Bode's since he had been the "defender of the faith".

Finally, there was the respective fame of the authors. Quite simply, Titius' translation and his Law are the only things that give him a real place in scientific history. Bode, on the other hand, eventually became a world-famous astronomer, founder and editor of the *Astronomische Jahrbuch*, responsible for many important observations, and even had an autobiography published about him in 1801 **(3.10)**. In short, Bode was a very important man. This can best be summed up by noting their respective author titles. Titius **(3.2)** was simply "der Naturlehre Professorn auf der Universität Wittenberg", whereas Bode **(3.8)** was "Königl: Astronom, Ritter des Preuss: Rothen Adler- und des Russisch: St.-Annen-Ordens zweiter Klasse, Mitglied der Akademien und Gesellschaften der Wissenschaften zu Berlin, Petersburg, London, Stockholm, Kopenhagen, Göttingen, München, Utrecht, Moskau, Verona, Hanau, Breslau &c:".

Despite this, it is unquestionable that the Law is Titius'. Perhaps this was his one good idea and Bode might have had many others; but the fact remains that it was Titius' idea and he deserves the credit for it, not Bode. Occasionally there have been voices objecting, such as Joseph Jérôme Le Français de Lalande (1732–1807) **(3.11)**, Jean Baptiste Biot (1774–1862) **(3.12)**, and the great natural scientist Baron Friedrich Wilhelm Heinrich Alexander von Humbolt (1769–1859) in his classic *Kosmos* **(3.13)**, but such complaints have accomplished little. The Law has mainly become known as the Bode Law with little attention being given to its true author.

The situation reminds one of the disagreements often heard today between the editors of letter journals and authors. Does the ultimate scientific credit for an idea at least partially depend on who you are and where you publish? In this case, it most certainly did.[†]

[†] A recent short biography on Bode with bibliographic sources can be found in ref. **(3.14)**.

CHAPTER 4

Success with Uranus and the Asteroids

FOR many years after the formulation of the Law no great stock was placed in it. This fact later moved Bode to declare **(4.1)**, "It is noteworthy that as yet no mention has ever appeared of this progression in the astronomical work of foreigners. Only German astronomers have mentioned it after I [*sic*] drew attention to it in my astronomical writings"[†] **(4.2)**.[‡]

This situation existed when in 1781 William Herschel (1738–1822) discovered "a curious either nebulous star or perhaps a comet" **(4.3)** which upon further investigation he "found that it is a comet, for it has changed its place". Even though the object had a resolvable disk and no tail, it did not occur to Herschel that it would turn out to be the seventh planet Uranus (a name, incidentally, that Bode chose). Little wonder, for no new planet had been discovered for thousands of years[††] **(3.5)**.

In fact, it was Nevil Maskelyne (1732–1811), the Astronomer Royal, who first realized that Herschel might have discovered a new planet. Despite this, much effort still went into computing parabolic orbits, which were doomed to fail, of course. A few months later, Anders Jean Lexell (1740–84) became convinced that Herschel's object was a planet and he published the first circular orbit calculation **(4.4)**.

[†] "Bemerkenswerth ist es, dass noch nie in astronomischen Werken der Ausländer von dieser Progression die Rede gewesen. Bloss deutsche Astronomen haben solche vorgetragen, nachdem ich in meinen astronomischen Schriften zuerst darauf aufmerksam gemacht."

[‡] A translation into English of Bode's remark is found in ref. **(4.2)**.

[††] The stories of the discoveries of Uranus, the minor planets, and especially of Neptune, which we discuss in Chapters 4 and 5, are fascinating unto themselves. We will go into them as they refer to us, but I highly recommend Morton Grosser's book, *The Discovery of Neptune*, ref. **(3.5)**, for a thorough account.

16

It was only then that most astronomers realized that a new planet had been found.

There were problems in calculating a good orbit, however, since observations extended only over a few months. To overcome this, Bode started a search for old observations that might have been mistaken for a star and found two. When these were taken into account, Pater Placidus Fixlmillner (1721–91) found a mean distance for the new planet of 191·8254 **(4.5)** which could be compared with the Titius–Bode prediction of $4+3 \times 2^6 = 196$, a deviation of only 2%.

Suddenly a new meaning was available for the Law, for it had "predicted" what no one would have thought possible, a new planet. Bode now was entirely convinced of the Law, and thus he also believed that there had to be a planet between Mars and Jupiter.

Baron Francis Xaver von Zach (1754–1832), the court astronomer at Gotha, had also become convinced, and in 1787 he began an intermittent methodical search for the trans-Martian planet. In 1800 he hit upon the scheme of dividing the sky into twenty-four zones to be searched by twenty-four astronomers. But in January, 1801, just as the star maps were being sent out, one of the twenty-four, Giuseppe Piazzi (1746–1826), found an object which proved to be Ceres. Piazzi wrote letters to Bode[†] and Lalande saying it was a comet, only confessing in the letter to his friend Barnaba Oriani (1752–1832) that he thought it was a planet.

von Zach immediately accepted it as the new planet. Even though no one else had seen Ceres because it had approached too near the Sun, von Zach published his article, "On a long supposed, now probably discovered, new major planet of our solar system between Mars and Jupiter" **(4.6)**.

Not everyone was ready to accept a new planet, however. Note the case of the great philosopher Georg Wilhelm Friedrich Hegel (1770–1831). Soon after the mail reached von Zach, Hegel managed to publish his docent thesis, *Dissertatio philosophica de Orbitis Planetarum* **(4.7)**. Here Hegel "logically proved" that the number of planets

---

[†] Ref. **(4.1)**, p. 1. See also the translation into English on p. 180 of ref. **(4.2)**, where the discovery of Ceres by Piazzi is detailed.

could not exceed seven by adding a planet between Mars and Jupiter (shades of Rheticus and Kepler). He thus refuted those astronomers who were trying to find a new planet to satisfy a numerical progression. As part of this demonstration, Hegel emphasized that the philosophical progression, 1, 2, 3, 4, 9, 16, 27, would not need a new planet. This pleased Hegel, for he considered Kepler's philosophical geometric and musical harmonies to be more fundamental than Newton's abstract and technical calculations **(4.8)**.

Although a series of articles was published on Ceres, it could not be rediscovered because enough information could not be gleaned from Piazzi's observations to calculate an orbit. To do this, Karl Friedrich Gauss (1777–1855) had to develop a major, whole-new method of celestial mechanics, the method of "least squares". Armed with Gauss' prediction, von Zach and Heinrich Wilhelm Matthäus Olbers (1758–1840) were able to independently rediscover Ceres on the anniversary of its discovery. The mean distance of Ceres as calculated by Gauss was 27·67 as compared to the Titius–Bode distance of 28. The Law was confirmed! Kepler's planet was found!

But the planet was very small, and what was worse, a few months later Olbers discovered Pallas, which had a mean distance of 26·70. What to do with this one and how to save the Titius–Bode Law? Herschel expressed the problem well **(4.9)**. "There is a certain regularity in the arrangement of planetary orbits which has been pointed out by a very intelligent astronomer, so long ago as the year 1772;" (anyone ever heard of Titius?) "but this, by the admission of the two new stars into the order of planets, would be completely overturned; whereas, if they are of a different species, it may still remain established."

A solution was devised by Olbers who suggested that the small planets were the fragments of the disruption of the original "major" planet. This meant that there had to be more "minor" planets. When Karl Ludwig Harding (1765–1834) found Juno in 1804 and Olbers found Vesta in 1807, this explanation seemed valid (even though the total asteroid mass was much less than any planetary mass) and the doubters of the Law were put to flight. The Law was "true"!

18

CHAPTER 5

Neptune and Other Problems

AFTER the discovery of the minor planets, the Titius–Bode Law was held so firmly that it was something on which to base future planetary astronomy. There was only one nagging point, which had been noted but not pursued.

Vikarius (Johann Friedrich) Wurm (1760–1833) first touched on it in 1787 (**5.1**). [In fact, here Wurm first put the Law in the form of eqn. (1.1).] Wurm pointed out that the Law should really predict a distance of $5\frac{1}{2}$ instead of 4 for Mercury, a distance half-way between its real distance and that to Venus. In other words, the index in eqn. (1.1) should be $n = -1$, the next term of the progression, with others possible.

Gauss later took this point further. In a letter to von Zach (**5.2**), he argued that if the Law really is correct as given, then there should be an infinite number of allowed orbits between Mercury and Venus, all the orbits to $n = -\infty$. Otherwise, one has the problem of explaining why the first planet has the essentially *ad hoc* relationship to the rest of the planets. In writing this letter, Gauss was apparently ignorant of Wurm's work of 15 years' standing, a situation that moved Wurm to remind people of his priority (**5.3**).

All this was really refocusing attention on the old question of a missing planet between Mercury and Venus. But even after all the new planets had been discovered, most people were not too concerned and were quite happy to leave the question alone.

A propos to this same subject, it should be noted that Bailey (**5.4**) has recently suggested that the Moon is just this missing planet; that it was in an unstable orbit; and that it became captured by the Earth. At the very least, this thesis is plausible since the geological evidence from the Apollo missions indicates a separate lunar origin.

Returning to the 1820's, except for the missing planet(s) question,

19

all was well with the Law and Bode could happily refer to it with a firm belief in its validity **(3.8)**. However, during this period the first hint of what would be a new trouble appeared in the strange behavior of the orbit of Uranus. In 1790 Jean Baptiste Joseph Delambre (1749–1822) had constructed tables for Uranus' orbit that satisfied the then known "ancient" observations (those that had been mistaken for a star) and also the modern ones. By 1820 these tables disagreed with observations.

Alexis Bouvard (1767–1843) attempted to correct this with new tables, but he could not get any tables to agree with all the observations. Bouvard's solution was to disregard all the "ancient" observations of renowned astronomers whose other stellar observations were known to be accurate. This was indefensible, and for it Bouvard suffered scathing attacks by Friedrich Wilhelm Bessel (1784–1846) and Urbain Jean Joseph Leverrier (1811–77). But it did raise the question of what really was the solution.

Many hypotheses were tried, such as a variation in Newton's Law at distances as large as Uranus. Finally, though, attention focused on the possibility of their being another planet causing perturbations.

This idea, however, was not new. As early as 1758, in his prediction **(5.5)** of the 1759 return of (Edmund, 1656–1742) Halley's Comet, Alexis Claude Clairaut (1713–65) had pointed out that a planet at such large distances might not have been seen and could affect the comet's orbit. Such a conception was truly impressive since this was even before the discovery of Uranus. Later, Wurm **(5.1)** had calculated many trans-Uranus orbits from a modified Titius–Bode Law (see Chapter 6). Further, in 1802 Ludwig Wilhelm Gilbert (1769–1824) had speculated **(5.6)** that the aphelion distance (35·5 AU) of the 1759 (Halley's) Comet was associated with the trans-Uranus planet predicted by the Titius–Bode Law at 38·8 AU. In fact, because of the then recent discovery of Ceres, discussion of such a planet was so widespread that a tentative name had been given to it **(5.6)**, "Ophion".

However, since three decades had produced no new planet, the idea had evaporated; so that it was in a fresh climate that the concept arose again in the 1830's. Then the first to commit himself to the idea probably was the Rev. Dr. Thomas John Hussey, the rector of the church at Hayes, Kent, and an amateur astronomer. He did this in an 1834 letter **(5.7)** to the Astronomer Royal, George Biddel Airy (1801–92), who promptly and rudely discouraged him.

In the following year observations on the orbit of Halley's Comet convinced both **(5.8)** Jean Élix Benjamin Valz (1787–1861) and **(5.9)** Friedrich Bernhard Gottfried Nicolai (1793–1846) that there was a perturbing force caused by an unknown planet, just as Clairaut had speculated nearly a century earlier **(5.5)**. They both assumed that the distance of this planet was near 38 AU, as predicted by the Titius–Bode Law. Indeed, Niccolo Cacciatore (1780–1841) and Louis François Wartmann (1793–1864) found and lost moving stars which they suspected of being the missing planet. The latter said that[†] "the new planet must needs lie at about double the distance of Uranus from the Sun," **(5.10)**. Clearly, the fortunes of the Titius–Bode Law had prospered.

The story of the solution of this problem is one of astronomy's most fascinating tales. As Morton Grosser **(3.5)** has vividly described it, John Couch Adams (1819–92) predicted the location in the sky of a new planet only to have his prediction ignored and buried by the Astronomer Royal, Airy. A similar calculation by Leverrier also met apathy. But he finally convinced Johann Gottfried Galle (1812–1910) to search for it. On Galle's first night of observation, September 23, 1846, he found it right where predicted.

What is interesting to us is that both Adams and Leverrier did their calculations assuming that the Titius–Bode Law was valid so that they used semi-major axes of 37·25 **(5.11)** and 36·15 **(5.12)**, respectively. But in 1847 Sears Cook Walker (1805–53) found an "ancient" observation by Lalande that proved to be Neptune. This allowed him to calculate a reliable orbit, and he found that the only

---

† "Si l'on admet la loi de progression des distances au Soleil, suivie approximativement par les autres planètes, et qui n'est qu'empirique, *il faudrait que cette nouvelle planète fût à une distance du Soleil à peu près double de celle d'Uranus*, exprimée par le chiffre 388, celle de la Terre au Soleil étant 10; ce qui donnerait environ 243 ans pour la durée de sa révolution."

parameter that agreed with Adams and Leverrier was the true longitude on January 1, 1847. In particular, Walker's semi-major axis was 30·25 AU and not a higher number as would come from the Titius–Bode Law (5.13).

The reason, as Lyttleton (5.14) has discussed in his analysis of the discovery, is that a not-too-bad value for the distance to Neptune could be compensated for by a change in assumed mass and still could give a good prediction for mean longitude over a short period. But, whatever else could be said, the Titius–Bode Law had broken down badly, a situation that certainly was not improved when Pluto was discovered (5.15, 5.16) in 1929–30 at a distance of 39·5 AU instead of the Law's 77·2 AU.

No longer was the Law something to base future predictions on. Instead, it became something of a black sheep. So much seemed to be in it, but yet it broke down badly in the end. Was it an approximate law or a law that should be put in a different form? Could a theory be made for it? These were questions that after 1847 had to be asked.

A final anecdote concerns the "discovery" of the planet "Vulcan". In 1859, after studying the orbit of Mercury, Leverrier (5.17) came to the conclusion that (given the correct mass for Venus) an error in the assumed motion of Mercury's perihelion could only be accounted for if there existed some unknown planet or planets between Mercury and the Sun. Immediately thereafter, M. Lescarbault, a physician at Orgères, Eure-et-Loir, France, stated that he had observed what he thought was the new planet on March 26, 1859. (Leverrier later calculated Lescarbault's elements as $r = 0.143$ AU, $\tau = 19^{d}17^{h}$.)

The full history of the "discovery" of Vulcan and of the four-decade search to rediscover it (including such events as the sighting of a flight of cranes in front of the Sun's disk) is at times hilarious and can be read elsewhere[†] (5.18–5.22). Let it just be said that because of the other new planets and Leverrier's fame belief in Vulcan was at first widespread. However, even though as late as 1879 a modified Titius–Bode Law was proposed to incorporate Vulcan (5.23), belief in Vulcan

[†] The "discovery" is reported in refs. (5.18) and (5.19). Later histories are given in refs. (5.20–5.22).

slowly waned. Despite this, Leverrier believed in the planet(s) until the end of his life[†] **(5.24)**, for there was no other way he could account for the anomalous *advance of Mercury's perihelion by 31" of arc per century.*

With hindsight we can see that Leverrier was a scientist whose past success with an idea blinded him to another possibility later. Recall that the deviation of Uranus' orbit could have been due to (1) the existence of another planet, or (2) a deviation from Newton's Law. For Uranus, Leverrier was correct in choosing the first possibility and discovered Neptune. So he automatically and mistakenly ruled out the second choice for Mercury's orbit—the choice dictated, as we now know, by the general theory of relativity. (And lest it be said that poor observations kept Vulcan from being discovered, an unsuccessful search was made for Vulcan during the March 7, 1970, solar eclipse.)

[†] "La conséquence est très-claire. Il existe dans les environs de Mercure, entre la planète et le Soleil sans doute, une matière jusqu'ici inconnue. Consiste-t-elle en une ou plusieurs petites planètes ou bien en des astéroïdes ou même en des poussières cosmiques? La théorie ne peut prononcer à cet égard. À de nombreuses reprises, des observateurs dignes de foi ont déclaré avoir été témoins du passage d'une petite planète sur le Soleil; mais, on n'est parvenu à rien coordonner à ce sujet."

CHAPTER 6

Early Modifications of the Law

IN 1785 von Zach pointed out in a "sealed letter"[†] to Bode that there was a large deviation from the Law for the outer planets if one fitted the first two terms of eqn. (1.1) to the distances of Mercury and Venus. To rectify this the Law's distance between Mercury and Venus needed to be smaller.

Two years later Wurm (5.1) accomplished this when he proposed what were the first modifications and further applications of the Titius–Bode Law. He noted that a better fit could be obtained for the entire planetary system if one considered the more general equation

$$r_n = a + b \times 2^n, \qquad n = -\infty, 0, 1, 2, \ldots, \tag{6.1}$$

and set $a = 0.387$ and $b = 0.293$.

More importantly, Wurm was the first to take the point of view that if the Law has a significant dynamical origin, one would expect that a related Law exists for the satellite systems of the major planets. We can comprehend this by realizing that either in a nebular theory (where the central major planet condensations would also have nebulas) or in a theory where solid body interactions were the origin of the Law, the satellite systems would present similar conditions for the dynamical mechanism to enforce the Law's pattern. In fact, the existence of a generalized Law that encompasses the satellite systems could be considered a test of the Law's validity.

Wurm found that reasonably good Laws for the then known satellites of Jupiter and Saturn could be given by

$$r_n = (3.0) + (3.0)2^n, \qquad \text{Jupiter}$$
$$r_n = (4.5) + (1.6)2^n, \qquad \text{Saturn} \tag{6.2}$$
$$n = -\infty, 0, 1, 2, \ldots,$$

[†] "versiegelten Zettel". See p. 171 of ref. (3.6).

where the unit of distance was the radii of the parent planet and where the fourth and sixth orbits were empty in the case of Saturn.

The next modification was made in 1802 by Gilbert (5.6) who realized that the Law need not be tied to a geometric progression ratio of 2. (Unfortunately, many others have also/still not realized this.) Thus, he placed the Law in the form

$$r_n = a + bc^n,$$
$$n = -\infty, 0, 1, 2, \ldots. \tag{6.3}$$

In units of Saturn's radius, Gilbert applied eqn. (6.3) to that system with the constants $a = 3 \cdot 08$, $b = 0 \cdot 872$, and $c = 2 \cdot 08$.

A similar observation was made in 1828 (6.1) by James Challis (1803–82), who suffered a bitter fate elsewhere in this story.[†] Challis applied eqn. (6.3) to Jupiter, Saturn, and Uranus, mistakenly thinking he was the first to investigate satellite systems. (This happened to more than one author.) With the rough agreement he obtained, he made the following conclusions: (a) the Law holds, except that Saturn seems to have a two-fold progression, (b) "c" is likely an integral or half integral number, (c) "a/b" might be a ratio of simple integers, and (d) the deviation from the Law is probably due to the masses and mutual actions of the member bodies. His fourth conclusion was due to the commensurabilities of the Jupiter satellites.

Soon after the discovery of Neptune a continuing interest in problems of this kind was manifested by Daniel Kirkwood (1814–95). Kirkwood was fascinated with regularities in the solar system and ways to observe them. This fascination was certainly rewarded, for it led him to discover the gaps in the asteroid belt that bear his name (6.2, 6.3).

His first proposal on this problem was "Kirkwood's Analogy"

[†] Challis directed the Cambridge search for Neptune. Due to the influence of Airy and "lack of confidence" in Adams, he delayed the search for a year. When he did start, he decided to make a wide-area search, instead of concentrating around the position Adams had pinpointed. Even so, Challis actually had located the planet twice in the first 4 days, but he did not check his data.

(6.4), which caused quite a controversy for some years (6.5–6.7). This Analogy states that for two successive planets,

$$K = T^2/D^3,\qquad(6.4)$$

where K is a constant. For any planet, T is the ratio of its sidereal day to its sidereal year and D is the "diameter of its sphere of attraction" (the distance over which its gravitational potential would dominate during a total planetary conjunction). He arrived at this conclusion by discussing the way condensation would cause planets to spin positively and by numerical testing.

It was from the above view of primordial pairs of planetary rings[†] (6.8) that Kirkwood was led to the view that a better Law of Planetary Distances was "The differences of radii of gyration of the primitive rings form a geometric series," (6.7). In equation form,

$$\frac{x_{n+1}-x_n}{x_n-x_{n-1}} = K',\qquad(6.5a)$$

$$x_n \equiv \left[\frac{r_n^2+r_{n+1}^2}{2}\right]^{1/2},\qquad(6.5b)$$

where r_n is, as before, the distance to the nth planet and K' is a constant. Kirkwood applied this Law to the planetary and satellite systems, obtaining an interesting rough agreement in general and a close agreement in Saturn's case. In Saturn radii assuming satellites between Rhea and Titan and between Hyperion and Iapetus, eqn. (6.5) produced $(x_{n+1}-x_n)$ of 2·13, 5·08, 12·13, and 28·96, in almost exact agreement with $K' = 2·385$. Kirkwood further noted that groups of satellites seemed to have periods near those of the planets from Mars outward.

What is fascinating about all of this is that Kirkwood's ideas, such as rings, pairs of rings and/or planets, relations to rotation periods, mean distances of periodic comets (6.9), etc., were all precursors to similar ideas of the next century.

[†] Kirkwood mentions that a paper by Steven Alexander, ref. (6.8), induced him to reinvestigate planetary pairs of rings.

Kirkwood was not alone in his search for "harmonies" in the solar system. For instance, in 1849 Benjamin Peirce (1809–80) came to the conclusion **(6.10)** that the phyllotactic series 1–2, 1–3, 2–5, 3–8, and 5–13 could be found not only *in the arrangement of buds and other organs of plant growth*, but also in the ratios of the orbital periods of planets. What Peirce had noted, in this rather odd way, is the tendency towards commensurability between orbiting bodies. (This will be discussed in Chapters 8 and 10.) In fact, as one might have expected from his later discovery of the gaps in the asteroid belts, Kirkwood eventually came to a similar conclusion himself **(6.11)**.

During the same period, Steven Alexander (1806–83) completed **(6.12, 6.13)** the work he had started many years before **(6.8)**. Alexander proposed that the consecutive ratios of planetary distances should be related to either the one-half power, the three-quarters power, or to the first power of the ratio of the distances to Saturn and Jupiter. He used that ratio as his basis since his calculations indicated Saturn and Jupiter would have the same moment of inertia at the time of the condensation of the planets from the solar nebula.

Pliny Earle Chase (1820–86) proposed a planetary distance Law of the form **(6.14)**

$$r_n = \frac{\pi}{32}(1 + n\pi),$$

$$n = 1, 2, 3, 5, 9, 17, 33, 65, 97,$$

(6.6)

that was later slightly modified **(6.15, 6.16)**. This series was motivated by hypothesizing a vibrating nebula that had a nodal structure with respect to distance. Of course this series is strictly *ad hoc*, since the numbers n were picked arbitrarily for a fit and there was no dynamical calculation to obtain such a series.

However, one can see that all the ideas of the preceding three paragraphs [which are collectively discussed in refs. **(6.15, 6.16)**] show that there existed a deeply ingrained belief in a harmonic or regular structure to the solar system. Certainly this had been true previously, and as we shall see in the later chapters of this book, it is still true today.

At the turn of this century, there was a series of eqn. (6.3) type Laws proposed for the satellite system of Saturn by George Frederick Chambers (1841–1915) **(6.17)**, Karl P. T. Bohlin (1860–1939) **(6.18)**, and Carl V. Ludwig Charlier (1862–1934) **(6.19)**. The fascinating question with this system, of course, is what to do with the rings. Are they to be considered as (a) separate orbit(s), or should the edges be considered as orbits? Bohlin considered the rings' midpoint as the $n = -\infty$ orbit; but Charlier, using the planet radii as the normalization, gave a Law of

$$r_n = (1\cdot5) + (1\cdot6)(1\cdot5)^n, \qquad \text{Saturn} \qquad (6.7)$$

and associated the ring with the orbits from $n = -\infty$ to -1.

In other words, Charlier, being unaware of Wurm's prior work **(5.1, 5.3)**, accepted Gauss' **(5.2)** observation on the missing distances for $n = [-1, -\infty)$ and used it as the reason for the ring's existence. The fact that Mercury has a large eccentricity was interpreted as a sign that it really is the $n = -\infty$ planet and that it was formed from a large fraction of the original solar "ring". An interpretation of Charlier's picture, then, is that the existence of Mercury vs. the ring of Saturn is due to Mercury not violating the (Édouard, 1820–83) Roche Limit **(6.20, 6.21)** and thus being stable against tidal breakup (see Chapter 10).

Of course, since the Titius–Bode Law was an object of such interest during the nineteenth century, there were numerous other works which discussed the Law in addition to those we have already mentioned. Many of them proposed modified Laws based on the types of numerologies we have already investigated. In refs. **(6.22–6.29)** some of the most interesting articles are listed. The reader who wants to pursue this point even further and to take up related questions is referred to the bibliography in ref. **(6.30)**.

All of these modifications, however, lacked a fine degree of agreement with observation, often missing by quite a few percent, and there was a fair amount of arbitrariness in their parameterizations. In addition, except for Kirkwood, the modified Laws were still mainly bound to the original form, something from which many of the theories we shall discuss will also suffer.

The psychological hold of the Law on astronomy has been such that people have always tended to regard its original form as the one on which to base theories. Closer examination shows this to be a weak argument. If there is truth in the Law, the original form should be thought more likely to be a *good first guess, but certainly not necessarily the best guess* to which to refer theories! The need to search for possibly better phenomenological formulations should have been more clear.

In the next chapter we will look at two formulations which did succeed in such a search and hence provide a better way to judge the content of the Law.

Before going on, however, it is amusing to note one other "Analogy" in the solar system that was observed by James Utting in 1823 (**6.31**). It was reported elsewhere (**6.32**) and caused a short spate of excitement (**6.32–6.34**), but then was never heard from again. Apparently, just by playing with the numbers, Mr. Utting observed that all of the planets revolving about the Sun and all of the satellites revolving around their respective mother planets satisfied amazingly well relationships of the form

$$v_n(r_n)^{1/2} = \text{const.}, \tag{6.8}$$

where v_n is the orbital velocity of the nth body.

Unfortunately for Mr. Utting, though, it was quickly pointed out (**6.33, 6.34**) that he had been scooped by 204 years. To see this one simply writes down the period of revolution of the nth body (τ_n) as

$$\tau_n = 2\pi r_n/v_n, \tag{6.9}$$

and combines eqs. (6.8) and (6.9) to give Kepler's Third Law (**2.5**):

$$\tau_n^2 = (\text{const.})\, r_n^3. \tag{6.10}$$

To Mr. Utting's defense, it should be pointed out that he had only done what Kepler had done; i.e. he had tried to get an empirical fit to the data, and he was successful. However, as Utting's work was not noticed to be equivalent to Kepler's, it must also be observed that the refereeing system of the *Philosophical Magazine* in 1823 certainly was not quite as picky as that of, say, the *Physical Review Letters* in 1971.

CHAPTER 7

Blagg–Richardson Formulation

THE modifications of the Titius–Bode Law that we discussed in Chapter 6 were mainly of the form of eqn. (6.3) and none were ever in exceptionally good agreement with all four systems of the Sun. This was changed by two formulations done in this century which were impressively accurate and quite similar, although formulated independently.

The first was done by Miss Mary Adela Blagg (1858–1944) **(7.1)** in 1913. Miss Blagg started off by looking at the log of the distances and attempting to find what was the best "average" difference in the log of the distances. After much trial and error, she came to the conclusion that the best Law for the solar system is given by a progression in 1·7275 (*not* 2·0) (see Fig. 7.1) multiplied by a periodic function that represents the deviation from the exact progression in each case. To be more explicit,

$$r_n = A(1{\cdot}7275)^n y$$
$$\equiv A(1{\cdot}7275)^n\,[B + f(\alpha + n\beta)]. \tag{7.1}$$

A and B are positive numerical constants and α and β are angular constants. f is a periodic function of 2π radians that ranges between 0 and $+1$, which is shown in Fig. 7.2. For comparison with the figure Miss Blagg gave an unnormalized off-origin harmonic analysis for f of

$$f = 0{\cdot}4594 + 0{\cdot}168\cos 2(\theta - 60{\cdot}4°) + 0{\cdot}053\cos 4(\theta - 77{\cdot}2°) + \ldots$$
$$+\, 0{\cdot}396\cos(\theta - 27{\cdot}4°) + 0{\cdot}062\cos 3(\theta - 28{\cdot}1°) \tag{7.2}$$
$$+\, 0{\cdot}009\cos 5(\theta - 22°) + 0{\cdot}012\cos 7(\theta - 40{\cdot}4°) + \ldots$$
$$\simeq \frac{\cos\psi}{3 - \cos 2\psi} + \frac{1}{6 - 4\cos 2(\psi - 30°)}, \qquad \psi = \theta - 27{\cdot}5°. \tag{7.3}$$

For the planetary system[†] **(7.2)**, with the constants shown in Table

[†] The data on the planets and satellites is taken mainly from Allen, ref. **(7.2)**.

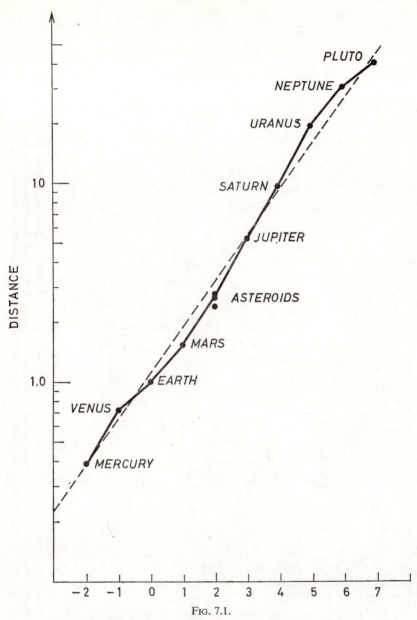

Fig. 7.1.

A semi-log plot of distance in astronomical units vs. ordinal number, for the planets. The straight line fit has a slope of 1·7275, which is the Blagg–Richardson geometric progression ratio.

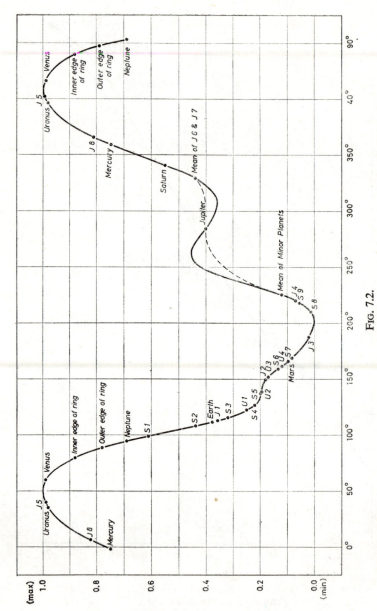

FIG. 7.2.

The f function in the Blagg formulation. It varies between 0 and +1, and is a periodic function of 360°. The dashed line is a suggested change which makes the two sides of the curve more symmetric. However, this probably depends on the correct choice of Saturn and Uranus constants (see text).

32

TABLE 7.1

The constants of the Blagg formulation of the Titius–Bode Law.

The α's are adjusted from the original paper by an increase of 2β for the planets and Jupiter and 4β for Saturn, to be consistent with eqn. (7.1) and Fig. 7.2.

System	A	B	α	β
Planets	0·4162	2·025	112·4°	56·6°
Jupiter	0·4523	1·852	113·0°	36·0°
Saturn	3·074	0·0071	118·0°	10·0°
Uranus	2·98	0·0805	125·7°	12·5°

7.1, this produced remarkable agreement with observation (see Table 7.2*a*).

Before proceeding, let us discuss the meaning of the constants. A is the unit of distance or normalization. In a certain sense, B is a measure of the deviation of eqn. (7.1) from an exact geometric progression. As B becomes large, the factor f becomes less and less important. Contrariwise, as B approaches zero, the deviation of eqn. (7.1) from an exact progression becomes larger. Specifically, the maximum and minimum allowed ratios of the distances of two successive planets are $1·7275(B+1)/B$ and $1·7275(B)/(B+1)$.

β is the step along the curve f that accounts for the deviation from an exact geometric progression in each particular case of two successive orbits. α is the starting point on curve f, but it is arbitrary up to a factor of $n\beta$ just as A depends on the assignment of n. That is to say, if we change our number assignment by n', this will not change the form of r_n, but it will change A by a factor of $(1·7275)^{n'}$ and change α by a factor of $n'\beta$.

Miss Blagg also looked at the satellite systems. Amazingly, she ultimately found that the best progression ratio for Jupiter was given by the same factor (1·7275) even though she originally started with quite different numbers. The deviations from a progression were given by the same function y with different values of the constants (see Table

TABLE 7.2

Comparison of the Blagg formulation with observation.

The bodies in parentheses were not known in 1913, when the original paper was written. Unoccupied and newly occupied distances have been calculated using Fig. 7.2. As seen in the Table, some of the calculated distances in the Saturn and Uranus systems are not very accurate. This is because the low values of B make them very sensitive to the exact form of Fig. 7.2.

Table 7.2a

Planet	n	Distance	Blagg Law
Mercury	−2	0·387	0·387
Venus	−1	0·723	0·723
Earth	0	1·000	1·000
Mars	1	1·524	1·524
Vesta ⎫		2·361 ⎫	
Juno ⎪	2	2·670 ⎬	2·67
Pallas ⎬		2·767 ⎪	
Ceres ⎭		2·767 ⎭	
Jupiter	3	5·203	5·200
Saturn	4	9·546	9·550
Uranus	5	19·20	19·23
Neptune	6	30·07	30·13
(Pluto)	7	(39·5)	41·8

Table 7.2b

Jupiter system	n	Distance	Blagg Law
J. V	−2	0·429	0·429
(J. XIII ?)	−1	(0·62 ?)	0·708
Io	0	1·000	1·000
Europa	1	1·592	1·592
Ganymede	2	2·539	2·541
Callisto	3	4·467	4·467
	4		9·26
	5		15·4
J. VI ⎫		27·25 ⎫	
J. VII ⎬	6	27·85 ⎬	27·54
(J. X) ⎭		(27·85) ⎭	
(J. XII) ⎫		(49·8) ⎫	
(J. XI) ⎪	7	(53·3) ⎪	55·46
J. VIII ⎬		55·7 ⎬	
(J. IX) ⎭		(56·2) ⎭	

34

Table 7.2c

Saturn system	n	Distance	Blagg Law
(Janus)	−3	(0·538)	0·54
Mimas	−2	0·630	0·629
Enceladus	−1	0·808	0·807
Tethys	0	1·000	1·000
Dione	1	1·281	1·279
Rhea	2	1·789	1·786
	3		2·97
Titan	4	4·149	4·140
Hyperion	5	5·034	5·023
	6		6·3
	7		6·65
	8		7·00
Iapetus	9	12·09	12·11
Phoebe	10	43·92	43·85

Table 7.2d

Uranus system	n	Distance	Blagg Law
(Miranda)	−2	(0·678)	0·64
	−1		0·77
Ariel	0	1·000	1·000
Umbriel	1	1·394	1·393
Titania	2	2·293	2·286
Oberon	3	3·058	3·055

7.1). Further, similar plots could be made for Saturn and Uranus. The comparison with observations is shown in Table 7.2.

Three decades later, Richardson (**7.3**) did a similar analysis and came to the conclusion that the distance should be of the form

$$r_n = (1\cdot728)^n \varrho_n(\theta_n),$$
$$\theta_n = n(4\pi/13) \equiv n\theta_1,$$

(7.4)

where ϱ_n is an oscillatory function of 2π. He came to this conclusion by studying a straight-line, semi-log fit for the planets and noting the oscillatory behavior (see Fig. 7.1). Looking along the straight line in Fig. 7.1 is in essence the same as looking across the $f = 1/2$ line in Fig. 7.2. (Note that $4\pi/13$ is $35\cdot4°$ as compared to Blagg's solar β of $56\cdot6°$.)

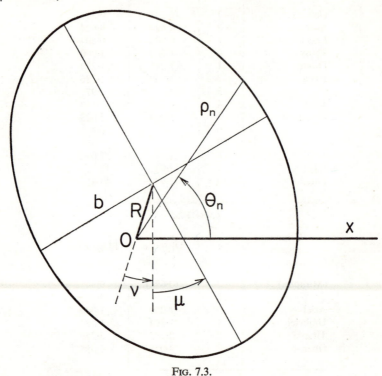

FIG. 7.3.

The "distribution ellipse" of the Richardson formulation.

Richardson chose to represent the distances ϱ_n as those from an off-centered origin to angularly varying points on a "distribution ellipse" (see Fig. 7.3). ϱ_n is given by

$$0 = [1 - e^2 \sin^2 (\theta_n - \mu)]\varrho_n^2 - R\{(2 - e^2) \sin (\theta_n + \nu)$$
$$- e^2 \sin [\theta_n - (2\mu + \nu)]\} \varrho_n + R^2[1 - e^2 \cos^2 (\mu + \nu)] - b^2, \quad (7.5)$$

where e is the eccentricity and the other constants are labeled in

Fig. 7.3. μ is similar to Blagg's α. Although θ_1 was kept a constant by Richardson, the parameter ν does give a little angular freedom. b, R, and e are related to A, B, and the ratio B/f of Blagg. For completeness we give the constants of Richardson (see Table 7.3) and the results he obtained from his formulation (see Table 7.4).

TABLE 7.3

The constants for the Richardson formulation of the Titius–Bode Law.
b *is given in miles, except for the planets.* θ_1 *is* $4\pi/13$.

	Planets	Jupiter	Saturn	Uranus
e^2	0·2352	0·2352	0·5170	0·5170
b	0·1900 AU	49,210	24,500	30,500
R/b	0·1588	0·2353	0·1355	0·472
μ	$\theta_1/3$	$-5\theta_1/12$	$\pi/2+\theta_1/24$	$-3\theta_1/4$
ν	$\theta_1/6$	$\theta_1/12$	$-\theta_1/24$	$3\theta_1/8$

As one could have expected, Richardson generally obtained the same good agreement as Blagg. However, for one satellite of Saturn there was an exception. Richardson's computed value for Mimas was too low by a factor of 1·66 (!), and Mimas is not a "small" satellite. What Richardson did differently was to number the satellites consecutively from Mimas to Iapetus, including a gap only between Iapetus and Phoebe. There were not the gaps that Blagg chose between Hyperion and Iapetus or even the long-postulated gap between Rhea and Titan. The Mimas results lend credence to Blagg's choice of gaps, always bearing in mind that many people would not have the outer satellites included in Titius–Bode Laws since they could have been captured.

However, Blagg's choice implies a rather great deviation from an exact geometric progression (B small). This might be justified in Saturn's case since Janus fits in. But in the case of Uranus, Miranda is off by 6%. Even though Miranda is small, this could mean that there is a need for better constants since there were only four satellites fitted to begin with. Perhaps a more rational fit would be with constants

TABLE 7.4

Comparison of Richardson's formulation with the then existing observations.

We have kept Richardson's original normalizations, which were in AU for the planetary system, and in miles for the satellite systems.

Table 7.4a

Planets	n	Distance	Richardson Law
Mercury	1	0·387 099	0·386 888
Venus	2	0·723 331	0·723 992
Earth	3	1·000 000	0·999 423
Mars	4	1·523 688	1·525 193
Ceres	5	2·767 303	2·869 509
Jupiter	6	5·202 803	5·193 472
Saturn	7	9·538 843	9·505 306
Uranus	8	19·190 978	19·210 380
Neptune	9	30·070 672	30·300 493
Pluto	10	39·457 43–	41·833 607

Table 7.4b

Jupiter system	n	Distance	Richardson Law
J. V	1	112,600	112,712
	2		
Io	3	261,800	261,137
Europa	4	416,600	415,537
Ganymede	5	664,200	664,196
Callisto	6	1,168,700	1,171,913
	7		
	8		
J. VI, J. VII, J. X	9	{ 7,114,000, 7,292,000, 7,350,000 }	7,609,243
J. XI, J. VIII, J. IX	10	{ 14,040,000, 14,600,000, 14,880,000 }	11,405,556

Table 7.4c

Saturn system	n	Distance	Richardson Law
Mimas	1	115,300	69,680
Enceladus	2	147,800	146,900
Tethys	3	183,000	182,538
Dione	4	234,400	234,860
Rhea	5	327,300	326,940
Titan	6	758,800	756,890
Hyperion	7	919,700	718,110
Iapetus	8	2,210,000	2,224,500
	9		
Phoebe	10	8,034,000	7,812,060

Table 7.4d

Uranus system	n	Distance	Richardson Law
	1		
Ariel	2	119,100	119,367
Umbriel	3	165,900	165,394
Titania	4	272,200	272,844
Oberon	5	364,000	364,334

something like $(B, \alpha, \beta) = (1 \cdot 04, 40°, -50°)$ with the suggested change in Fig. 7.2, and Miranda placed in the $n = -1$ orbit.

This is related to the question of whether the progression ratio, $1 \cdot 73$, which is valid for the heavy parent-body solar and Jupiter systems, has to be changed to a lower number for the lighter parent-body Saturn and Uranus systems. In these systems a pure geometric progression fit would yield a lower ratio than $1 \cdot 73$. Further, for them the Blagg periodic function f does not go through one cycle (2π radians). (The Richardson periodic function ϱ_n travels less than one cycle in the case of Uranus.) This is an indication that here lower progression ratios and/or perhaps larger angular steps from satellite to satellite might be in order. Later we will mention physical reasons why this might be true, but for now we just emphasize this point to keep it in mind.

In any event, however, for the systems where the number of objects is large, the fit is striking.

Four sets of comments are now in order.

(a) *Shape of the y function*

Although the shape of the Blagg y (or f) function is important from a theoretical point of view, what is significant from a phenomenological standpoint is that all systems fit the same smooth f function, as one would want for a valid "general" law. Clearly, however, a more definitive statistical analysis could be applied to the Richardson formulation or to the even–odd approximate representation of eqn. (7.3). Even so, we will discuss the original f function of Blagg because of the convenience in representing all systems.

(b) *New objects*

As Roy **(7.4)** has pointed out, many new objects have been found since Blagg's formulation was proposed and they should fit into the scheme if it is true. We have placed these new objects in parentheses in Table 7.2 and have calculated all the new and unoccupied distances from the Law. (Note, however, that since B is small for Saturn and Uranus, in certain regions the newly calculated distances are very sensitive to the form of f, and so are not certain beyond the first order.)

The first object is Pluto, which at 39·6 AU fits into the $n = 7$ slot which predicts 41·8 AU. Furthermore, for a trans-Pluto planet, $n = 8$ would predict 67·7. Schuette **(7.5)** seems to find a group of comets at about 80 AU aphelia. This is suggestive since it long ago has been noted **(5.1, 6.9)** that groups of comets associated with planets have aphelia somewhat farther than the planets' orbits.

For the satellites, Janus fits beautifully into the $n = -3$ slot of Saturn. Jupiter IX, XI, and XII fit into the $n = 7$ position along with

Jupiter VIII; and Jupiter X fits into the $n = 6$ position with Jupiter VI and VII. Uranus V (Miranda), at a distance of 0·678 on its scale, belongs in the $n = -2$ position. Finally, if the newly predicted **(7.6, 7.7)** very small satellite of Jupiter at 3·66 radii ($\equiv$ 0·62) exists, it would best fill the $n = -1$ Jupiter position of 0·708.

(c) *Goodness of fit*

As we have noted, for the most part there is a striking agreement with the Blagg–Richardson formulation. Even the "bad" objects deviate mainly by only a few percent.

Further, as Richardson pointed out, the "bad" objects are always "small" in the sense that they subtend a small solid angle at the center of the parent body when compared to the other satellites. In fact, for his formulation, Richardson plotted the curves of percentage deviation vs. satellite number and they had the same forms as the curves of solid angle subtended vs. satellite number. (The exception was Mimas, for which Blagg got a good fit.) Roughly speaking, good-fit satellites (planets) subtend a solid angle at the center of the parent body of 10^{-5} (10^{-8} to 10^{-10}) whereas bad-fit satellites (planets) subtend 10^{-8} (10^{-12}).

Finally, the deviating bodies tend to be in groups, like the asteroid belts and the outer Jupiter satellites. They also tend to be at places where the f curve is very steep. It was at these places that Blagg expected many small bodies and where, for example, the Saturn rings are found.[†]

(d) *Numerical observations*

In addition to his observations on the smallness of "bad" objects, Richardson noted other numerical generalities. These can be read in his article, but we mention the two that appear to be the most interest-

[†] These observations are in principle pertinent to the tidal and gravitational mechanisms discussed in Chapters 8, 10, and 13.

ing. First, for the three satellite systems, one can write

$$D = \frac{2}{5} \frac{b}{R} \sqrt{(ab)}, \tag{7.6}$$

where D is the diameter of the parent body and a and b are the semi-major and semi-minor axes of the distribution ellipse. Further, for all four systems there is the rough empirical equation

$$|\mu+\nu| = \sin^{-1}\left\{0.178\left(\frac{b}{R}\right) e^2(2-e^2)\right\}. \tag{7.7}$$

Equations (7.6) and (7.7) are good to a few percent.[†]

To conclude this chapter, we summarize the results of Blagg and Richardson.

1. A general Titius–Bode type Law can be obtained that fits all four systems of the Sun to a much better degree than the original formulations.
2. This Law is a geometric progression in 1·73 *(not 2)* multiplied by a periodic function.
3. With the progression in 1·73, there is no need for the first term of the original Titius–Bode Law, and a much better fit is clearly obtained for the planetary system.
4. The periodic function represents the deviation from a pure geometric progression. The relative deviation is not the same for the four systems, but to good accuracy it can be represented by the same mathematical function.

Given this better formulation of the Law, we can now investigate the dynamical significance of it.

[†] Richardson **(7.8)** also did a phenomenological analysis of the planetary masses.

CHAPTER 8

Evidence for the Law's Validity

RECENTLY analyses have been done on the statistical significance of apparent regularities in the solar system: to be explicit, on (a) commensurabilities in mean motion $(n) \propto (1/\text{period of orbit})$; (b) geometric progressions for planetary and satellite distances (Titius–Bode type Laws); and (c) resonance relations.

These regularities are all pertinent to the existence of a generalized Titius–Bode type Law. Since we have just seen the striking fits that are obtained with the Blagg and Richardson formulations of the Titius–Bode Law, a discussion of the above statistical analyses is in order. This will give us information on the significance that can be placed in the Law.

(a) *Commensurabilities in mean motion*

It has long been known that there are many pairs of bodies that have mean motions that are very nearly commensurable, that is, the ratios of their mean motions are very nearly the ratios of integers (i/j).

The most famous cases are in the Jupiter system where the mean motions of three of the four Galilean satellites[†] Io (n_1), Europa (n_2), and Ganymede (n_3) satisfy

$$\frac{n_2}{n_1} = \frac{1}{2} - 0 \cdot 001817,$$

$$\frac{n_3}{n_2} = \frac{1}{2} - 0 \cdot 003647,$$

(8.1)

and furthermore obey the relation discovered by Pierre Simon

[†] Galileo (Galilei) (1564–1642) discovered these first, new, *moving*, heavenly bodies in 1610, soon after starting to use the telescope.

Marquis de Laplace (1749–1827) **(8.1, 8.2)**

$$n_1 - 3n_2 + 2n_3 = 0 \qquad (8.2)$$

to nine significant figures (observational accuracy):

$$n_1 - 2n_2 = 0 \cdot 739\ 469\ 091,$$
$$n_2 - 2n_3 = 0 \cdot 739\ 469\ 092. \qquad (8.3)$$

According to Laplace this commensurability is exact (stable) **(8.1)**.

In addition to the above, there are many more close commensurabilities. Examples are Jupiter and Saturn (5/2), Tethys and Mimas as well as Dione and Enceladus (1/2), and Hyperion and Titan (3/4).

Roy and Ovenden **(8.3)** investigated the statistical significance of these commensurabilities. Their analysis, as modified by Goldreich **(8.4)**, is as follows: First, a limit must be set to i and j, for otherwise exact, but yet meaningless, commensurability could always be obtained. Arbitrarily this limit is set at 7. The forty-six pairs of objects ($p_1, \ldots, p_{46}$) in the solar system whose mean motions are within a factor of 7 of each other are then studied. For a given kth pair, the deviation from commensurability (ε) is defined as

$$\varepsilon = \left| \frac{n'_k}{n_k} - \frac{i}{j} \right|. \qquad (8.4)$$

For this same commensurability, n'_k/n_k is most closely bounded above and below by some (i/j) and (i'/j'). Thus, the probability that n'_k/n_k is randomly within ε_0 of one of these ratios can be taken as

$$P_{\varepsilon_0}(k) = \frac{2\varepsilon_0}{\left(\dfrac{i}{j} - \dfrac{i'}{j'} \right)}. \qquad (8.5)$$

This calculation can be repeated for all forty-six pairs of objects. If, apart from the particular distribution of fractions, the pairs show no preference for commensurability, then the expectation value (E_{ε_0}) for the number of near commensurabilities with $\varepsilon \leqslant \varepsilon_0$ is

$$E_{\varepsilon_0} = 2092\varepsilon_0. \qquad (8.6)$$

The results are shown in Table 8.1 and clearly indicate a tendency towards commensurability.

TABLE 8.1

The expected number of "near" (within ε_0) commensurabilities (E_{ε_0}) for a random distribution of mean motions, and the number of "near" commensurabilities observed (N_{ε_0}) for various values of ε_0.

0·0119 is one-half of the difference between the closest two fractions, i.e. $0·0119 = (1/6-1/7)/2$.

ε_0	0·0119	0·0089	0·0059	0·0030	0·0015
E_{ε_0}	24·9	18·6	12·3	6·3	3·1
N_{ε_0}	33	26	20	12	6

(b) *Geometric progressions*

Dermott **(8.5)** has investigated the statistical significance of geometric progressions. However, because of the results on commensurabilities, he chose to consider a geometric progression in orbital periods, i.e. the progression

$$\tau_n = \tau_0 A^n. \tag{8.7}$$

Because of Kepler's Third Law, this would correspond to a distance law of

$$r_n = r_0(A^{2/3})^n. \tag{8.8}$$

He restricted A^2 to be either the ratio or the square of the ratio of integers—again, because of the commensurability data.

Dermott studied only the "regular" (small inclination and eccentricity) inner satellites of the major planets. He ignored the planetary system for two reasons. The first was that, as Alfvén **(8.6)** has pointed out, it might be more profitable to study satellite systems since we know three of them and only one planetary system.

A second minor reason was that ter Haar and Cameron **(8.7)** suggest a distance progression of 1·89 as the best for the planets. But this clearly breaks down since it has to exclude *both* the Earth and Pluto in the numerical sequence to obtain agreement. The number 1·89 is actually a theoretical number from the von Weizsäcker Theory (see Chapter 14), so this is not a good excuse to ignore the planetary

45

system. We have seen that the far lower progression of 1·7275 (lower still from 2·0) is a very good fit to the planets.

Even with these restrictions, Dermott did find a statistical validity to geometric progressions in the satellite systems with the values of A^2 shown in Table 8.2. Note that Dermott's Jupiter A^2 is the closest to the Blagg fit. This is understandable since this system is the closest (largest B) to a pure geometric progression (see Tables 7.1–7.4).

TABLE 8.2

The geometric progressions in period (A) found by Dermott (8.5)
for the regular satellites.

$A^{2/3}$ is the corresponding distance progression. The Blagg–Richardson progression is given for comparison.

System	A^2	$A^{2/3}$
Jupiter	4	1·59
Saturn	2	1·26
Uranus	3	1·44
Blagg–Richardson Progression	$5·155 = (2·27)^2$	1·7275

We can also see from Tables 7.1–7.4 and Fig. 7.1 that the planetary system is the "least evolved system" in that it is the closest to the pure Blagg–Richardson progression. Since

$$[(1·7275)^{3/2}]^2 = 5·155 = (2·27)^2, \qquad (8.9)$$

the pure Blagg progression would produce at best a poor fifth-order commensurability between objects two orbits from each other.

Finally, we mention that Dermott also pointed out that with his values of A, the ratios of τ_0 to the rotational period of the mother planets for the satellite systems are 1·076 (J), 1·079 (S), and 1·076 (U). He took this as an indication of other couplings.

(c) *Resonance relations*

A resonance relation between bound objects is of the form

$$\sum_{j=1}^{N} \alpha_{ij}\omega_j = S_i, \qquad (8.10)$$

where the α's are integers and the $\omega_j \propto n_j$ are the angular frequencies of the N objects. If $N-1$ relations can be formed where $S_i = 0$, then the system is in a maximum resonance.

Molchanov **(8.8)** showed that there exist circumstances under which a maximum resonance structure is inevitable in a dissipative harmonic medium *if* enough time is given. (As would be expected, lower α's correspond to a more stable resonant structure.)

Molchanov then constructed a set of complete resonance relations for the four systems of the Sun. From the low values of the α_{ij}'s that he obtained he concluded that the solar system is in a resonant state. However, although Molchanov's relations are indicative, Backus **(8.9)**, Henon **(8.10)**, and Dermott **(8.11)** came to the conclusion that Molchanov's low α_{ij} values are still compatible with a random distribution of ω's, a conclusion that Molchanov disputed **(8.12)**.

Thus, we are left with the facts that commensurabilities and geometric progressions exist to a statistically significant degree with possibly a tendency towards "total commensurability" (total resonance structure). However, all of these facts are embedded in the Blagg–Richardson formalism. For example, the commensurabilities, like those of the Galilean satellites of Jupiter, are magnificently represented; but they surely do not come from a geometric progression. As mentioned above, the closest commensurability in the Blagg–Richardson progression is the bad fifth-order commensurability between bodies two orbits apart.

We are thus led to our **(8.13)**

FIRST CONCLUSION. In the Titius–Bode Law, the geometric progression and the y (evolution) function are of separate origin. The y function in eqn. (7.1) is a representation of commensurabilities and/or a tendency towards commensurabilities. It has its origin in the phenomenon which created the commensurabilities. The geometric progression represents a separate phenomenon and origin[†].

Previously, people have sometimes investigated Titius–Bode geometric progressions in terms of commensurabilities, for example in **(8.5)** and **(8.8)**. Most often they have concentrated solely on the geometric progressions.

However, with our conclusion one should not attempt to look for the origin of the Titius–Bode geometric progression in the origin of commensurabilities or vice versa. The problem of finding a theory to explain the Titius–Bode Law is now split into two parts.

We should point out, however, that our separation of the physical content of the Law into the same two parts as given in the parameterizations of Blagg and Richardson is not strictly iron clad. It could be argued that the tendency towards commensurabilities might not be totally contained in the y function (or Richardson's ϱ function). However, because of the points we have presented, we feel justified in continuing on the basis that our First Conclusion is substantially correct, if not absolutely.

[†] It is amusing to speculate **(8.13)** if the progression ratio 1·73 really represents $3^{1/2}$ or $\pi^{1/2}$.

CHAPTER 9

Origin of the Solar System and Hoyle's Theory

WE ARE now in a position to ask for the reasons behind the Titius–Bode Law, more particularly the Blagg–Richardson formulation of it. But first we must touch upon the related and more general question of the origin of the solar system. We will not include a review of this field since reviews already exist (8.7, 9.1, 9.2), and the interested reader is referred there. However, we do note that solar system theories can be divided into two classes: (1) nebular theories and (2) encounter–capture theories. Recently, most discussion has centered around nebular theories and, indeed, these seem to be better suited for explaining Titius–Bode type Laws, as we shall see. Nevertheless, there are some non-nebular theories which try to explain the Titius–Bode Law and we shall mention them later.

That being said, we will consider the solar system as having had a nebular origin, specifically, that of the theory of Hoyle (9.3, 9.4). The reason is that even though Hoyle did not discuss the Titius–Bode Law one way or the other, his theory still has the advantage that it offers an explanation for most of the observed peculiarities of the solar system. Thus, even if it is not the totally correct theory, it is a convenient and good reference with which to organize our knowledge of the solar system.

Hoyle describes the following evolution:

When the proto-sun had gravitated to a radius of approximately 0·2 AU (3×10^{13} cm), it reached a rotational instability against further contraction. The instability was caused by the angular momentum in the gas cloud

$$\mathscr{L} = 4 \times 10^{51} \text{ g cm}^2 \text{ sec}^{-1}, \tag{9.1}$$

a number obtained by saying that the original cloud had 1 atom/cm³ and had an angular velocity on the order of that of the galaxy as a whole ($\sim 10^{-15}$ sec^{-1}).

After noting that almost all of the solar system's angular momentum has been deposited in the "planetary material", eqn. (9.1) is also in agreement with the presently observed angular momentum of the solar system, *providing* that the present planetary material is only about 10% of that in the original disk and also that most of the lost material was from the regions of Uranus and Neptune. This viewpoint can be defended by observing that Uranus and Neptune have concentrations of the common non-metals (C, N, O, Ne) which are about 10^2 higher than that in the Sun. (Jupiter and Saturn do not need to have a large correction to their relative H–He concentrations.) Increasing the masses of Uranus and Neptune by this factor of 100 overwhelms the Jupiter and Saturn contributions to the mass and angular momentum of the solar system. Further, it means that the total expelled mass was about 3000 Earth masses and that the angular momentum of the planetary material, after it had spread out into a disk but before the H and He had evaporated from its outer regions, was in agreement with the value for $\mathcal{L}$ in eqn. (9.1).

To get the radius of the Sun when it reached instability (R_I), Hoyle took θ as a constant and set[†]

$$\mathcal{L} \simeq \tfrac{1}{10} MR^2\theta \equiv M\langle R^2\rangle\theta. \tag{9.2}$$

(The right-hand side of eqn. (9.2) is a factor of 4 smaller than it would be for a uniformly rotating solid sphere, implying a slightly condensed core.) Combining eqns. (9.1) and (9.2) with the condition of equal centrifugal and gravitational forces,

$$R\theta^2 = \frac{GM}{R^2}, \tag{9.3}$$

yields

$$R_I \simeq 3\times10^{12} \text{ cm} = 0.2 \text{ AU}. \tag{9.4}$$

At the time of ejection, the Sun's internal temperature of $\sim 10^3$ °K was able to ionize enough hydrogen (only about one atom in 10^7 was necessary) to allow a strong magnetic coupling between the inner condensing proto-Sun and the expelled material. This was the mecha-

† In this book we will reserve the use of M, as well $M_\odot$, for the mass of the Sun.

nism for the transfer of angular momentum from the Sun to the expelled matter. The field acted as a brake[†] (9.5, 9.6) on the Sun with the lines of force looking something like spokes on a wheel that were being wound like a clock spring. The field lines attached the inner "hub" (proto-Sun) to the outer wheel (the expelled matter).

This transfer of angular momentum expanded the expelled matter into a disk. Angular momentum was continuously transferred by the magnetic field to the inner radius of the disk (which probably remained near the same distance, R_I), and the angular momentum was then further transferred to the outer reaches of the disk by turbulent friction.

Hoyle showed that during the transfer of angular momentum the field windings were first stored in the disk (about ten of them). But as the tangential component of the Sun's field at the surface (H_t) became equal to the radial component (H_r), the inward magnetic pressure on the surface of the Sun, which exists for $H_t > H_r$, allowed the field windings to be stored in the Sun even though this ordinarily would have been forbidden in the highly conducting medium of the Sun. The storage was possible only because the Sun also had a deep convective zone which allowed the lines of force to diffuse quickly enough in the interior.

Thus, Hoyle gave an explanation of the very striking fact that the Sun has very little angular momentum compared with the planets. The theory also predicts that dwarf stars in general have planetary systems. In contradistinction, it further predicts that massive stars will be rotating rapidly (as is observed) since the lack of a deep convective zone in these stars will prevent the magnetic brake from working effectively.

The rate of transfer of angular momentum can be calculated by noting that when the radial and tangential components of the Sun's field become equal, the stress opposing the rotation is

$$F = (4\pi R^2)\frac{H_r^2}{4\pi} = R^2 H_r^2. \qquad (9.5)$$

[†] Such a magnetic brake was first discussed by Alfvén (9.5).

Thus, the torque on the Sun is of order

$$\Gamma \sim R^3 H_r^2. \tag{9.6}$$

But due to the conservation of flux,

$$R^2 H_r^2 = \text{const.} \tag{9.7}$$

This shows that the transfer of angular momentum is largest when R is largest. So, we can approximately say that

$$\mathcal{L} \simeq \tau\Gamma(R=R_I) = \tau R_I^3 H_r^2 \tag{9.8}$$

for some contraction time τ. If we take τ to be the (Hermann Ludwig Ferdinand von, 1821–94) Helmholtz–(William Thomson Lord, 1824–1907) Kelvin contraction time[†]

$$\tau \simeq 10^{14} \text{ sec}, \tag{9.9}$$

and then

$$H_r \sim 1 \text{ gauss}, \tag{9.10}$$

which is quite reasonable.

With regards to the energy transferred by the angular momentum, first note that from eqns. (9.2) and (9.3) the energy of rotation at instability ($R = R_I$) is

$$E_\theta = \frac{1}{2} M\langle R^2\rangle\dot\theta^2 = \frac{1}{20} R_I^2 M\dot\theta^2$$

$$= \frac{1}{20} \frac{GM}{R_I} = 5\times10^{45} \text{ ergs.} \tag{9.11}$$

But by the virial theorem, the amount of kinetic energy (E_D) that could be transferred to a still captured disk is bounded by one-half of the disk's potential energy at the time of ejection, i.e.

$$E_D < \frac{1}{2} \frac{mMG}{R_I^2} \simeq \frac{1}{200} \frac{M^2G}{R_I^2}. \tag{9.12}$$

[†] Chandrasekhar (9.7) contains an excellent discussion of the physics of the Helmholtz–Kelvin contraction. Also included is a fine set of historical and modern references on Helmholtz's proposal (and Kelvin's answer) that this contraction gives the age and explains the energy of the Sun. (Nuclear energy sources, of course, were not then known.)

Since eqn. (9.12) is down by a factor of 10 from eqn. (9.1), most of the transferred energy had to be placed in the magnetic field windings of the Sun. For a solar field of 1 gauss at $R = R_I$, the original energy in the field was about 10^{36} ergs. Thus, the field had to be increased by a factor of $\sim 10^5$ to obtain a magnetic energy of 5×10^{45} ergs. This meant that in about 10^{14} seconds, 10^5 windings had to be stored in the Sun, or about one relative winding every 30 years, a figure that Hoyle felt was plausible.

Finally, as the disk expanded, the materials with high melting points (the metals and silicates found in the terrestrial planets) condensed out first. The total mass of these elements was small because of their relatively small solar abundance. The outer major planets were made of materials with low melting points, such as water, ammonia, and the common non-metals. The planetary material comprised only about 10% of the original disk material, and after about 10^7 years the bulk of the disk evaporated.

The satellite systems formed in roughly the same way. However, there was not an effective brake mechanism. Thus, one should expect that the satellites contain a larger fraction of the parent planet's mass and are relatively closer together, and that the parent planets retain fast rotation periods, all of which are observed.

With this description, we divide the history of the planetary system into three periods.

PERIOD I. *The disk period.* The period before the non-planetary constituents of the disk had evaporated.

PERIOD II. *The aggregation period.* The period when the condensed rocks and ices slowly came together to form the bigger planets.

PERIOD III. *The planet period.* The period in which the planets have existed as single bodies, with little further accretion, all evolution being due to direct gravitational or tidal interactions.

The satellite systems can be viewed as having undergone the same three periods with perhaps uncertainty in how to distinguish Periods I and II.

CHAPTER 10

Periods of the Law's Creation

WE NOW ask a major question: In which of the three periods did the phenomena which created the two portions of the Titius–Bode Law occur?

(a) *The geometric progression*

We can rule out Period III producing the origin of the geometric progression for the following reason. There are only two ways in which Period III could have had the origin: through direct gravitational interaction or through tidal interaction. However, the well-known theorem of Siméon Denis Poisson (1781–1840) rules out the first possibility while tidal theory rules out the second.

Poisson's Theorem[†] **(10.1–10.4)** states that in the expressions for the semi-major axes, there are no secular terms due to gravitational interactions between satellites in the first and second orders of perturbation theory. In essence, this means that a random "stable" distribution of planetary distances could not have evolved into a significantly different set of distances via point gravitational forces. Later we will come back to this important theorem and discuss its implications further.

In this respect, we digress momentarily to mention that Hills **(10.5)** has taken an opposing view on the possibility of a gravitational origin to the geometric progression. He has proposed that at the beginning of Period III, soon after conglomeration, there could have

---

[†] Following an observation by Laplace, Joseph Louis Comte de Lagrange (1736–1813) proved the theory in first order **(10.2)**. Actually, a complete proof for all semi-major axes in second order was first given by François Félix Tisserand (1845–96) **(10.3)**. Also see ref. **(10.4)**.

been a random, dynamically unstable configuration which evolved via point gravitational interactions into the present stable configuration. He numerically calculated the evolution of model planet systems and his results gave evidence for this hypothesis, as well as for a tendency towards commensurabilities in the dynamical relaxation.

For the reasons presented in this chapter, however, our viewpoint is that this evolution is more properly associated with the y function of Blagg. Then the amount of evolution needed would be much smaller since it would only have to be from the geometric progression distance. [See the discussion under (b) *The y (evolution) function.*] So, keeping this interesting idea in mind, we defer a more detailed discussion of Hills' work until Chapter 13.

Returning to our main line of thought, the second possibility (an origin of the geometric progression in Period III from tidal interactions) is ruled out by a similar argument. Since the work of Sir George Howard Darwin (1845–1912) **(10.6)**, tidal theory has pointed to the impossibility of producing a significant change in the planetary distances (say a factor of 4). The argument does not hold as well for the satellite systems. However, since the geometric progression is clearly the dominant term in the planetary system, as well as in the Jupiter system, we must reject a tidal origin. Otherwise, tidal theory would have to account for the universal evolution into our non-commensurate geometric progression of (1·7275).

In a similar manner we can rule out Period II. Here we conceive of a cylindrically symmetric disk of rocks and ices slowly conglomerating. While the system is totally symmetric there will be no tidal or gravitational evolution towards commensurability.

However, this is true even after symmetry is quitted. Consider, for example, the planetary system. To first approximation the largest deviation from symmetry would be in the proto-Jupiter ring. But it is impossible to think of a gravitational effect from Jupiter and a cylindrically symmetrical disk which would produce a non-commensurable geometric progression, especially in a period on the order of 10^7 years.

We are left with Period I. This, in fact, seems the most reasonable.

In Period I fluid dynamics would be important and the cylindrical symmetry inherent in a geometric progression is more understandable. Further, we now see that the geometric progression is by itself evidence for a nebular origin to the solar system since our discussion has shown the difficulty of explaining it in terms of point gravitational inter-actions. Thus we have the

SECOND CONCLUSION. The geometric progression in the Titius–Bode Law originated during Period I and is a manifestation of some fluid dynamical and/or electromagnetic process, i.e. not a purely gravitational or tidal mechanism.

(b) The y (evolution) function

Contrary to paragraph (a) we can immediately rule out Period I because there is no way that a tendency towards commensurability would arise in a cylindrically symmetric system. This leaves Periods II and III.

Within certain limits it has been shown that commensurabilities, once formed, are stable, both under gravitational disruption **(10.7)** and tidal disruption **(8.4)**. This led Roy and Ovenden **(10.7)** to pro-pose a gravitational origin to the observed commensurability tendency and it led Goldreich **(8.4)** to propose a tidal origin.

Here the Poisson Theorem is not as applicable to a discussion of a gravitational origin because the Theorem only applies to secular terms and is not valid in third order[†] **(10.8)**. This is important because now we are dealing with many-body systems evolving through much shorter distances (only the deviation from the geometric progression). These facts and the Mirror Theorem **(10.7)** lend credence to the pos-sibility that direct gravitational interactions are important.

This is one of the reasons why we consider Hills' work **(10.5)** to indicate that it might have been possible to have an evolution from a *geometric progression distance* by means of point gravitational

[†] See the discussions by (Dirk, 1902–66) Brouwer and Clemence **(10.8)**.

interactions. For a total evolution mechanism by means of gravitational interactions, one can question, among other things, whether the solar nebula could have (a) formed the planets in very unstable random configurations which then could (b) evolve past gravitational-tidal commensurate stabilities to the present configuration. Again these objections are not as strong for the smaller evolution pictured here. (See Chapter 13 for a detailed discussion of Hills' work.)

Continuing, if one considers a tidal mechanism, the Darwin (10.6) objections do not hold for the y-function for somewhat the same reason; we are now discussing only the deviation from the geometric progression instead of from the entire distance.

However, this question hinges entirely on the rate of change of mean motion ($n \equiv \omega$) due to tidal disruption:

$$\frac{dn}{dt} = -\frac{27}{4} n^2 \left(\frac{m}{M}\right) \left(\frac{a}{R}\right)^5 \frac{1}{\left(1+\frac{19\mu}{2g\varrho a}\right)Q}, \tag{10.1}$$

where m is the satellite's mass; M, g, a, μ, and ϱ are the planet's mass, surface gravity, radius, rigidity, and density; and

$$Q = \frac{2\pi E_0}{-\oint \left(\frac{dE}{dt}\right) dt} \tag{10.2}$$

is the ratio of the maximum energy stored in the planet's tidal distortion to the integral over one cycle of the rate of energy dissipated. Since the quantity multiplying Q in the denominator of eqn. (10.1) can be taken as nearly 1 for the major planets, Q is the crux of the whole problem. It depends on such things as the planetary atmospheres, which would lose energy by turbulence, and the interior structure of the planets.

Goldreich (8.4) proposed a lower bound on Q by asking what is the lowest constant Q which would not have allowed the satellites close to Jupiter and Saturn (Io and Mimas) to be at the planets' radii 4×10^9 years ago. (He reduced the number for Io by a factor of 5 to 7·5 because it is in at least one stable commensurability.) Because of his belief that there had been tidal evolution, Goldreich felt that the satellites had to have evolved significantly so that the lower bounds are near the upper bounds. Thus, Goldreich proposed that

$$Q(J) \simeq (1 \sim 2) \times 10^5,$$
$$Q(S) \simeq (6 \sim 7) \times 10^4. \tag{10.3}$$

Goldreich then said that this could explain all the commensurabilities since driving the inner satellites out would also drive out the outer satellites, as they would be successively locked in.

However, Dermott **(10.9)** argued that Goldreich's Q's could not be taken seriously because any satellite which evolved from the surface of the major planets would have had to evolve through the radius (R_B) where its mean motion equaled the rotation period of the planet (and hence a radius where tidal action would cease). Furthermore, the satellite would have to have not violated the Roche Limit (R_L)[†] **(6.20, 6.21)**

$$\frac{r}{a} > (2 \cdot 52) \left(\frac{\varrho \text{ (planet)}}{\varrho \text{ (satellite)}} \right)^{1/3} \equiv \frac{R_L}{a}, \tag{10.4}$$

or else the satellite would have been disrupted by tidal action. Since there is no knowledge of the initial satellite densities for Jupiter and Uranus, where $R_L < R_B$, and since Dermott also questioned whether the atmospheric turbulence dissipation mechanism for Saturn could produce a small enough Q, Dermott doubted the Q bounds and so a tidal evolution, as had Jeffreys **(10.10)** earlier. Hence, Dermott proposed that the commensurabilities were a condition of formation, i.e. from our Period II or earlier.

Dermott's objections were admittedly non-rigorous so he certainly did not destroy Goldreich's theory. Furthermore, we still feel that Goldreich's conclusions are more plausible, even if not conclusive. This is for two reasons. First, in our picture **(8.13)** the satellites would not have to evolve from distances near the surface of the mother planet, only from the geometric progression distance. This would mean

[†] The original constant of Roche for the right-hand side of eqn. (10.4) was 10, not 2·52, obtained for an elliptical fluid satellite.

that

$$\frac{\Delta n}{n_0} = 1 - \left(\frac{r_{0n}}{r_n}\right)^{3/2}, \tag{10.5}$$

where r_{0n} is the geometric progression best fit.

One must realize that in terms of tidal action alone eqn. (10.5) would imply a tidal contraction for some planets; so perhaps this supports a gravity theory or means that one should consider the evolution from a geometric progression that bounds the planet distances from below. But in any event, eqn. (10.5) should give an order of magnitude number.

If we calculate eqn. (10.5) for Io, Mimas, and Ariel, either from the Blagg formalism (with the suggested changes for the Uranus fit) or from the Richardson formalism, we find the evolution is of order 20% (see Table 10.1). But Q's as high as $\sim 10^6$ (J) and $\sim 10^5$ (S, U) could produce the changes of Table 10.1, exhibiting that a tidal mechanism is easier to satisfy from our outlook.

TABLE 10.1

The fractional changes in mean motions corresponding to an evolution of the satellites to their present distances from the pure geometric progression distance.

For the Blagg formalism, the changes suggested in Chapter 7 for Uranus were used.

	Io (J)	Mimas (S)	Ariel (U)
Blagg formulation	−0·14	0·09	0·21
Richardson formulation	−0·09	0·23	0·14

To sum up, we can favor Period III and rule out Period II because the former lasts about 5×10^9 years whereas the latter lasts only 10^7 years. For the strengths of the mechanisms involved, 10^7 years is much too short a span to account for an evolution of the needed size. For certain of the bodies even 5×10^9 years would be too short a time to evolve a random or total scale distance. But since we say that the distances of evolution are only from some geometrical progression distance, the evolution distances we conceive of are compatible with the strengths of the mechanisms.

There is also a second reason for placing a tidal or gravitational mechanism in Period III. It is based on the following observation[†] **(8.13, 10.11)**.

> OBSERVATION. In the solar system commensurabilities between two solid-body satellites are stable, but commensurabilities between a satellite and the small bodies of a disk system are not necessarily bound. In fact, some are destructive resonances.

We have mentioned the proofs of commensurate stability in satellite objects and have seen them observed in the planets and satellites, not to forget the Jupiter–Trojan asteroids (1/1) commensurability.

However, consider the (1/2) commensurability between Mimas and the (Giovanni Domenico, 1625–1712) Cassini Division in Saturn's rings and the (1/4, 1/3, 2/5, 3/7, 1/2) commensurabilities between Jupiter and the Kirkwood gaps in the asteroids. These are examples of the second phenomenon and have been shown theoretically to be unstable.

This was done by Jürgen Moser **(10.12)**, who studied the stability conditions of the canonical differential equation system of celestial mechanics. When applied to the Newtonian three-body problem in the solar system (the Sun, Jupiter, and an asteroid; or Saturn, a satellite like Mimas, and a particle in the Saturn rings), then an unstable resonance is set up when

$$\frac{n_2}{n_3} = \frac{i}{j}, \quad k \equiv |i-j| \leqslant 4, \tag{10.6}$$

where i, j, and k are integers, and $n_3(n_2)$ is the mean motion of an asteroid or ring particle (Jupiter or a satellite like Mimas).

Since 98% of the mean motions of the asteroids have n_3 such that

[†] A similar observation was made earlier by Hagihara **(10.11)**.

$$\frac{1}{4} \leqslant \frac{n_2}{n_3} \leqslant \frac{1}{2}, \tag{10.7}$$

Moser explained exactly why those particular Kirkwood gaps mentioned previously appear in the asteroids and not others (for instance, $2/7$, where $k = 5$).

Thus, we know why these commensurabilities are vacant, they are unstable. If they had been stable the rings would have been in the division and the asteroids would have been in the gaps.[†]

But during Period II we would have this type of system, a first-order symmetrical disk with the beginnings of solid bodies. So, during Period II there is not any tendency to form commensurabilities. In fact, there is just the opposite, a tendency to move away from commensurabilities.

Thus, for two major reasons, we are left with the result that

THIRD CONCLUSION. The y (evolution) function was produced during Period III by tidal or point gravitational interactions.[‡]

As we have now placed the origin of the y function in Period III and ascribed its origin to tidal or point gravitational forces, it remains to explain the classical Titius–Bode geometric progression, whose origin we placed in Period I.

[†] The objection can be raised that Moser's calculation dealt only with circular orbits, not elliptical ones. With this in mind, Jefferys (10.13) proposed that the gaps are due to collisions among asteroids with *elliptical* orbits that have been captured in stable tidal commensurabilities. However, in some statistical sense we still view Moser's calculation favorably. This is because it predicted just and only just the correct asteroid gaps and because in the rings of Saturn, where the particle orbits should all be essentially circular, the predicted Cassini Division is indeed found.

[‡] For a general bibliography on current research into both point and tidal types of celestial mechanics problems, see the Report of the Committee on Celestial Mechanics of the I.A.U., in the latest (10.14) and preceding volumes of the *Transactions*.

CHAPTER 11

Search for a Geometric Progression Theory

IN THE next few chapters we will describe theories that have been proposed to explain the geometric progression of the Titius–Bode Law or other related "laws". In point of fact, in the Titius–Bode context these theories consider the Law a simple progression and not a progression multiplied by what we call the "evolution" function. Further, as we have mentioned, too much attention has been focused on the original geometric progression ratio of 2.

Be this as it may, we will describe these theories with the position that we are searching for an explanation to one part of the Law, the progression part. We will keep the viewpoint that there is another part of the Law, the "evolution" part, whose origin was discussed in Chapter 10.

Progression theories can roughly be categorized as either (a) electromagnetic, (b) gravitational, or (c) nebular theories, even though many of them have characteristics that could place them in more than one group. In the next chapter we will describe the electromagnetic theories and later proceed to the others.

Unfortunately, many of the theories that we will review are "theories" and "Laws" in large part due to assumptions or arguments of possibility rather than by calculations that predict definite and unambiguous results (including normalizations). Too often the Titius–Bode Law has been discussed in terms of some mechanism which could conceivably produce the Law if all other physical effects could be ignored. Sad to say, that is not the type of physical universe with which we are dealing.

Thus, one should be aware of the fact that any claim that a particular theory of the solar system is correct since it explains the Titius–Bode Law is not valid because, as of yet, no theory has properly explained this progression. What can be said of some theories is that they are compatible with and indicative of a Titius–Bode Law, but not very much more. That this is true will become quite clear in the next three chapters.

On the other side of the coin, though, the theories we will look at are often general theories of the solar system which are attempting to describe much more than just the Titius–Bode Law. This should be remembered since the mechanisms they propose are sometimes relevant in describing some of the other phenomena in the solar system, even if not in explaining the Titius–Bode Law.

To keep one's perspective, the reader should remember that here we are specifically investigating the Titius–Bode Law. Thus, in the next three chapters we will only review in detail those theories that have proposed a mechanism for the Titius–Bode Law and, in particular, will concentrate on the mechanisms involved. This means that we will not discuss some quite important theories of the solar system there, e.g. Hoyle's theory (9.4). The omitted theories have been left out simply because they did not propose a specific origin for the Titius–Bode Law.

Contrariwise, some of the theories we do discuss are clearly incompatible with information about the solar system that has nothing to do with the Titius–Bode Law. Nevertheless, we still include them. By doing this, in addition to clarifying those theories that are difficult to follow or hard to locate, our discussion can be complete and serve as a catalogue of ideas on the subject.

With regards to this last point, since our discussion in the next three chapters is rather detailed, we summarize here the ideas contained in them. Then the reader can easily find those parts which are of most interest to him.

Chapter 12. Electromagnetic Theories

We begin with the electromagnetic theory of Birkeland (A). Birkeland's idea was that the Sun emitted positive ions which had limiting cyclotron radii due to the Sun's magnetic moment. These radii give the progression Law of distances. We now know that the Sun's magnetic field is not large enough to capture ions in such orbits but the theory was important since it was the first to emphasize the importance of electromagnetic phenomena in the solar system.

Berlage (B) investigated whether a postulated strong, solar, electric field could create spherical shells of space charge that would create the Law by preferential condensation at those distances. Berlage abandoned these attempts because he did not see how the satellite systems could have had the fields necessary to create their Laws. But both here and in the field of nebular theories (see Chapter 14) his ideas were ahead of their time.

Alfvén (C) pictured a captured gas cloud which, upon falling towards the Sun, became ionized. By a magnetic brake mechanism the Sun's angular momentum was transferred to the ionized gas. Through this transfer of angular momentum to the massive ions Alfvén predicted a mass–distance Law. Alfvén's work was significant not so much in formulating a viable mass–distance Law, which it did not, but in its emphasis on a magnetic brake mechanism to slow down the solar rotation and on the importance of electromagnetic phenomena in the solar system.

Chapter 13. Gravitational Theories

We begin by discussing Schmidt's (A) theory, which assumes that the planetary material at one time captured a sea of meteorites. The mechanism he proposed for the origin of the Law was a conglomeration process in which the proto-planets, in competition with one another, aggregated meteorites on the basis of differences in angular momentum per unit mass. Among its defects, the theory needs an *ad hoc* additional process for the division of the Law into two regions (terrestrial and major planets). Further, the asteroid belt is quite mysterious, as in his theory it should contain a mass intermediate between the terrestrial and major planets.

Egyed (B) based his theory on the old Dirac hypothesis of a changing value of the gravitational constant with time. Then the Sun would have expelled matter during the early periods of its history to keep centrifugal and gravitational forces in balance. Independent of the theory's weaknesses, direct evidence can rule out G varying in time like Dirac postulated.

Woolfson (C) postulated that a stellar encounter with a massive star (~ 100 M) could have produced the Law. This was because approximately evenly spaced, traveling, tidal waves on the surface of the Sun would be pulled off by the passing star. However, he himself gave up on the attempt to obtain a valid, quantitative theory.

Pendred's and Williams' (D) model is that of a captured gas cloud falling by gravitational collapse onto the Sun. Basically by considering only this collapse they, obtain a mass and angular momentum distribution. After postulating a "gravitational reach" relationship, i.e. the limits of domain of one particular planet, they could predict a final planetary mass and distance Law. But we will see that this is rather arbitrary and unconvincing.

We next come to the theory of Hills (E) that we have mentioned already. By studying the dynamical gravitational evolution of model, point, planetary systems on a computer Hills came to the conclusion that the planets could have dynamically relaxed to the present distances from some initial, arbitrary distribution, this evolution showing a tendency towards commensurabilities. Even though there are a number of questions and objections that can be raised, this theory, which the reader will remember is in disagreement with our conclusions, is very interesting. In Chapter 15 we will mention ways in which this type of calculation can be extended to help clarify if the idea is viable and also to see whether it is more applicable only to the evolution function (as we would have it) or if it is indeed more applicable to the entire Titius–Bode Law.

Finally, we mention the just published theory of Dole (F) who, in a computer model calculation, inserts planetary "nuclei" into a primordial nebula of dust and gas. A particular nucleus is allowed to gravitationally sweep up the dust in a region whose size is determined by the nucleus' "random" initial orbital parameters. If, at the end of dust accumulation, the planet has reached a certain "critical" size, it can then start to sweep up gas and become a giant planet. When the first

65

planet has stopped growing, a new nucleus is inserted and the process continues until no more dust is left. Results are obtained which are roughly "solar system like", but there are a number of assumptions and dynamical simplifications in the theory. As a result, the idea, although an interesting one, must be viewed with reservations. One should notice that the end of Dole's model is the starting point of Hills' model. This raises the question of how Dole's planets would evolve under real gravitational and tidal evolution.

Chapter 14. Nebular Theories

Here we start off by reviewing the pioneering theory of Berlage (A), who looked at the equation of equilibrium of a rotating, gas nebula under the influence of gravity and internal gas pressure. By trying to minimize viscous energy loss in the system and obtaining from this a density maxima, Berlage obtained various forms of the Law depending on what kind of temperature and chemical weight distributions he used. Those that he did use turned out to be incorrect and he did not consider turbulence, but he deserves credit for being the first to try this approach.

We next discuss the theory of von Weizsäcker (B) which, during the past three decades, has aroused the most interest among scientists. von Weizsäcker's theory was the first to use the type of chemical composition for the solar nebula we know is correct, and it also was the first to recognize the importance of turbulence in the nebula. von Weizsäcker proposed that viscous forces from differential rotation about the Sun caused turbulence in the nebula. The turbulence in the nebula formed into a pattern of a whirlpool, vortex cells arranged regularly within rings. In this manner the energy loss due to friction could be minimized. This resulted in a prediction of a Titius–Bode Law since the condensation would preferentially take place at the interfaces of the rings and these interfaces are related in distance by the number of vortices in a ring.

This original theory was modified by a number of authors (C) who used the theory of turbulence to show that the sizes of the eddies that would exist in a turbulent nebula were of the order of magnitude desired. The final modifications showed that there would be a distribution of eddy sizes according to the Kolmogoroff turbulence spectrum, the largest of the eddies in this spectrum being consistent with von Weizsäcker's prediction. Further, it was to this eddy that most of the viscous energy would be transferred. The main questions remaining were if the largest eddies would dominate the form of the turbulence pattern and if the eddies, and even the turbulence itself, were long enough lived to begin with.

An opposing view was taken by Kuiper (D) who did not accept the idea that the largest eddies would dominate the form of the turbulence pattern over the rest of the Kolmogoroff spectrum. This and the problem of having long-enough-lived vortices induced him to propose that there really was a mass–distance Law. The mechanism came from the assumption that most of the nebula was originally near the Roche Limit for tidal instability. Then, random turbulence could create a high enough pressure to allow huge proto-planet gas spheres to become stable against disruption. The Roche Limit criteria for adjacent pairs of gaseous proto-planets at a particular radius yielded Kuiper's mass–distance Law. However, in reality, to get a reasonable agreement with the observed data, Kuiper had to arbitrarily modify the prediction he had made on questionable grounds in the first place. Further, Nakano, in proposing his own nebular theory, pointed out that Kuiper's gas cloud should not be able to contract rapidly enough to allow planets to be created.

CHAPTER 12

Electromagnetic Theories

A. Birkeland's Theory

The first "modern" astrophysical theory of the Titius–Bode Law was proposed by Olaf Kristian Birkeland (1867–1917) in 1912 **(12.1)**. Birkeland observed that since the Sun has a magnetic moment, particular positive ions emitted with a definite energy could spiral out towards a given final cyclotron radius that depends on the mass and charge of the ions. Birkeland proposed that in this way concentrations of certain ions would build up at certain distinct radii providing greater condensation possibilities to form the planets.

For two main reasons the theory could not stand in its original form. First, it turns out that the value of the Sun's field is not nearly strong enough to produce limiting radii within the present solar system. Also, according to this idea the chemical composition of all the planets should be quite different instead of being in two main groups, the terrestrial and the major planets.

The theory was important, however, for it was the first to introduce electromagnetic phenomena into the problem of the origin of the solar system. In this respect its pioneering work has not been widely enough recognized.

B. Berlage's Electromagnetic Theories

The history of the work of Hendrick Petrus Berlage, Jr. (1896–1968) is itself of interest. Not since Bode himself was anyone as continually occupied with the problem of the planetary distances. Berlage was convinced of the Law's validity, and for 40 years he retained an active interest in it.

His theories were forerunners of many others, but did not receive the attention or credit due to them. This was no doubt partially due to the way that Berlage presented his work. He wrote many papers on the subject, but they were often difficult to follow, especially since particular details of his theories often changed from paper to paper.

Berlage's initial efforts were aimed at finding an electromagnetic explanation to the Law. His first theory (12.2), in 1927, studied the possible effects of a solar electric field upon emitted charged particles, neglecting angular momentum. He assumed that the Sun emitted positive ions and negatively charged, solid particles that had condensed in the solar atmosphere.

Under these conditions, Berlage's model yielded a positively charged Sun surrounded by a negative cloud of solid particles. He calculated the equilibrium distance of a positively charged ion under the influence of solar gravity and the electric fields due to the positively charged Sun and the total surrounding space charge. That is, he found the solution of

$$GMm_p\mu + Q(r)e = Q_0 e. \tag{12.1}$$

Q_0, M, and $R = 0.00465$ AU are the charge, mass, and radius of the Sun. μ is the atomic number of ions of weight μm_p, and $Q(r)$ is the space charge within a sphere of radius r, which was calculated to be

$$Q(r) = K \ln \left[(1 + \sqrt{(1 - R/r)}) / (1 - \sqrt{(1 - R/r)}) \right], \tag{12.2}$$

where K is a constant.

In Berlage's model this produced a solution for the equilibrium radii of

$$r = R \cosh^2 \left\{ \frac{GMm_p}{2Ke} (\mu_0 - \mu) \right\} \tag{12.3}$$

$$\equiv \frac{1}{4} R[p^{(\mu_0 - \mu)} + 2 + p^{-(\mu_0 - \mu)}], \tag{12.4}$$

where μ_0 is that atomic isotope which would be bound at the surface of the Sun. Berlage took this to be 40 for calcium^{++} and hypothesized a normalization for p of 2 (the progression ratio).

Berlage then further hypothesized that (1) all the planet rings were formed a constant distance $d = 0.4$ AU further out from the ion rings (because of the addition of momentum to the growing planetesimals via radiation pressure), and (2) the planets from Neptune to Venus are represented by the atomic weights 25 to 32 whereas Mercury is 35, because isotopes 33 and 34 had not been discovered then. This allowed Berlage to explain the first-term anomaly since $0·00465 \times 2^6 = 0·3$ for Venus but only 1/8 of that for Mercury. Thus he had constructed the entire Titius–Bode Law:

$$r_n = R^{(38-\mu)} + d,$$
$$\mu_n = 35, 32, 31, 30 \ldots. \tag{12.5}$$

Berlage also tried to explain the satellite systems of the major planets and the masses of the planets with his model.

However, the theory failed for a number of reasons. First, while the article was in proof, isotopes 33 and 34 were discovered so that the anomaly could not be explained. Next, he neglected centrifugal forces and all the normalizations and the atomic number choices in his series were quite arbitrary. Most important, of course, we now know that no negative solid particles are emitted from the Sun to give the fields and ionizations necessary for the theory.

Berlage's next theory (12.3) accepted the evidence that the charged particles emitted by the Sun are positive ions and electrons and constructed a pair of equations for the two space charges, calculating the effects of the field from the space charge $Q(r)$ within a sphere of radius r, the radiation pressure $S_\pm$, and gravitation:

$$\pm Q(r)e + S_\pm - GMm_\pm = \frac{1}{2} m_\pm r^2 \frac{d}{dr} (v_\pm)^2. \tag{12.6}$$

For large distances the space charge velocities $v_\pm$ are equal and approach v_0. So, if n is the electric flux through a sphere of radius r, then when

$$c^2 + \frac{1}{4} \equiv a \equiv \frac{ne^2}{v_0^3} \left(\frac{1}{m_+} + \frac{1}{m_-} \right) > \frac{1}{4} \tag{12.7}$$

there exists a solution for $Q(r)$ which for some minimum r_0 is of the form

$$Q(r > r_0) = c_1 \sqrt{r} \left[c_2 \cos \left(c \ln \frac{r}{r_0} \right) + c_3 \sin \left(c \ln \frac{r}{r_0} \right) \right]$$
$$- c_1 c_2 r_0^{1/2} + Q(r_0), \tag{12.8}$$

where the c's are constants related to a.

Thus, the interpretation is that there are regions where $Q(r)$ is alternately positive and negative. Berlage called these nodes and anti-nodes of the field. The ions would tend to congregate at places of successive radii of minimum potential. From eqn. (12.8) the ratio of two successive such radii is

$$\frac{r_{n+1}}{r_n} = \exp \left(\frac{\pi}{c} \right) \equiv \exp \left(\frac{\pi}{\sqrt{(a - \frac{1}{4})}} \right), \tag{12.9}$$

i.e. a Titius–Bode type Law.

Berlage realized that his theory also had to explain the Laws for the satellite systems; and, in fact, he studied curves like Fig. 7.1 for these systems **(12.4)**. But after further consideration he came to the conclusion that it was impossible to have the planetary condensations emit the charged particles and set up the fields that he wanted. This was the reason that convinced him that he had to abandon this theory **(12.5)**.

C. Alfvén's Theories

With no indication of knowledge of the previous theories, in 1942 Hannes Alfvén began a series of three articles **(12.6)** in which he described the first of his electromagnetic theories on the origin of the solar system. Alfvén pictured the original solar dipole moment as being something like 10^5 times larger than at present, which meant that the ratio of the magnetic to gravitational forces on an ion was about 60,000.

This allowed him to propose a picture in which a cloud encountered by the Sun would fall under gravitational attraction until its kinetic energy equaled the ionization energy (V) at a distance of

$$r_i = \frac{GMm_p\mu}{V} = 14 \times 10^{13} \, [\mu/V(\text{eV})] \text{ cm.} \qquad (12.10)$$

Ignoring any original angular momentum of the cloud and given that the cloud is all ionized at the same distance from the Sun, a spherical mass distribution can be calculated,

$$\frac{dm(r)}{dr} = \frac{Mr_i}{2} r^{-2} \left[1 - \frac{r_i}{r}\right]^{-1/2}, \qquad (12.11)$$

which roughly agrees with that of the major planets.

Alfvén then considered the orbits that the ions would have. He argued the ions would slowly evolve to the disk via encounters, where they would be neutralized to form Kepler orbits with a calculated eccentricity of $e = 1/3$.

First consider the case of the region of neutralization moving in by encounters. The outer regions move in to a distance of $r' = (2/3)r_{\max}$ by encounters, at which time the particles are in circular orbits. However, because matter of lower angular momentum per unit mass was swept up, the final distance will be at r'', a little less than r'. Then a similar process begins at r'', meaning that successive rings will have

$$q \equiv \frac{r_{n+1}}{r_n} = \frac{r_n}{r_{n-1}} = \ldots = \text{const.} > 1. \qquad (12.12)$$

It is rather clear that this is a Titius–Bode Law, but Alfvén never even mentioned it. Instead he assumed the disk had a mass distribution of (h and s are constants)

$$\frac{dm}{dr} = hr^{s-1}. \qquad (12.13)$$

From eqns. (12.12) and (12.13) one obtains the conclusion that the ratio of the masses of two successive planets is

$$Q \equiv \frac{m_{n+1}}{m_n} = \int_{r_n}^{r_{n+1}} dm \, \bigg/ \int_{r_{n-1}}^{r_n} dm = q^s. \qquad (12.14)$$

But again Alfvén took another view.

Alfvén equated the angular momentum between r_n and r_{n+1} to the angular momentum of the planet, meaning that the expression which relates q to Q (for a condensation moving inward) is

$$\frac{Q\sqrt{q-1}}{Q-1} = \sqrt{\frac{2}{3}\left[1+\frac{\ln q}{2\ln Q}\right]}. \qquad (12.15)$$

Similarly, he obtained a q—Q equation for a condensation moving outward of

$$\frac{Q\sqrt{q-1}}{Q-1} = \sqrt{\left(\frac{3q}{4}\right)\left[1-\frac{\ln q}{2\ln Q}\right]}. \qquad (12.16)$$

Alfvén made the assertion that the outer major planets and inner satellites were due to a gas cloud condensing inward, whereas the inner terrestrial planets and the outer satellites were due to a separate encounter with a dust cloud which condensed outward. He then compared the q—Q values of these objects with the q—Q curves of eqns. (12.15) and (12.16). The *ad hoc* second cloud was necessary since the gas ions could never have penetrated as close to the Sun as the terrestrial planets. Alfvén felt that a dust cloud could reach the inner region before being decomposed by radiation.

This theory was harshly criticized by ter Haar and Cameron (**8.7**) for a number of reasons. (1) They disliked Alfvén's q—Q semi-log plot, feeling it gave a distorted view of an acceptable single curve on a log–log plot. (2) They pointed out that if eqn. (12.14) was correct (no matter sent to infinity), then the q—Q plot should really be a point, not two curves. (3) They criticized the condensation picture saying it is a hydrodynamic problem, not a simple question of Kepler orbits. (Objections were also raised about the size of Alfvén's original dipole, but a large original dipole is now widely accepted.)

Finally, ter Haar and Cameron objected strongly because Alfvén had the particles coming in from great distances without suffering collisions (long mean free path) and also had them suddenly ionized when they had enough energy to be ionized (short mean free path), both at the same time.

Other objections can be raised about the two separate clouds (gas and dust) that were needed arriving in the proper order and condensing in the opposite direction [not to mention the fact that Alfvén used $e = 1/3$ (2/3 contraction mechanism) to "explain" the Cassini Division in Saturn's rings]. Thus, the original Alfvén theory was ingenious, but it could not be accepted.

. Alfvén later produced a modified theory (12.7–12.10) in which he abandons his two cloud encounters for one nebula encounter at the time of solar formation. He uses the same ionization phenomenon as before, but this time he explicitly takes into account the fact that angular momentum will be transferred to the gas via the magnetic brake mechanism (9.5, 9.6). By setting the transferred angular momentum equal to the angular momentum of the planets (satellites), which in turn is related to the ionization energy, he finds that the important factor is the quantity

$$\xi^{-1} = \tau_i/\mathcal{O}. \tag{12.17}$$

τ_i is the Kepler orbital period corresponding to the ionization distance of the material that formed the satellite object. $\mathcal{O}$ is the average period of axial rotation that the parent body had during the time it transferred angular momentum to the satellite's material.

The planets and satellites are then placed into a number of groups of approximate ξ^{-1}: (A) Mars and the Moon; (B) the other terrestrial planets; (C) the major planets and the inner satellites of Jupiter and Saturn; (D) the outer satellites of Jupiter and Saturn, and Pluto.

Using the results of his previous theory, Alfvén predicts a mass concentration falling off going outward for complete ionization ($\xi^{-1} \gtrsim 50$), and a mass concentration becoming larger going outward for incomplete ionization ($\xi^{-1} \lesssim 50$). By choice, one or the other is roughly true in his groups, and he finally says that the groups came from four separate ionizations (clouds A, B, C, and D). The clouds came from the original nebula, when four different pieces neutralized as the nebula cooled, and then were re-ionized after they fell towards the Sun. Later clouds added satellites to existing bodies formed from the material of earlier clouds.

The theory, however, has many arbitrary assumptions: for instance,

74

that the angular momentum was transferred to the ionized "planetary" material and not to the entire nebula as in Hoyle's theory **(9.4)**; or that the number of clouds is four. (Why are there just four clouds and not five?) There are also large anomalies within the groups. Examples are his mass distribution of the outer Saturn satellites and the entire handling of the terrestrial planets, such as including Jupiter V in Cloud A.

Further, in the C cloud the planets decrease in mass outward (implying complete ionization). But this material was also to have given the material for the inner Jupiter and Saturn satellites, where the mass increases outward (implying incomplete ionization). The theory gets around this by saying that the ionized Jupiter material became neutralized when it arrived at the disk and only became partially re-ionized when it fell towards proto-Jupiter or proto-Saturn. But this whole sequence appears rather arbitrary. In fact, the whole theory is rather arbitrary, and one is left with the feeling that it is a parameterization to support a theory rather than vice versa.

Even if Alfvén's emphasis on the magnetic brake is very well taken, his extremely complicated mechanism puts in about as much as it gets out, and so it cannot be convincing. The Hoyle theory **(9.4)** is more agreeable in this respect.

Thus, although the importance of electromagnetic phenomena has been amply demonstrated, a convincing and valid explanation of the Titius–Bode Law has not been given that is due to an electromagnetic mechanism alone. It appears, as we have suggested, that one must concentrate on those mechanisms involving the entire hydrodynamics of the nebular disk and that one must beware of looking for the explanation of the entire structure of the solar system in terms of individual ions and electrons.

We shall return to this in Chapter 14.

CHAPTER 13

Gravitational Theories

IN THIS chapter we look at gravitational theories which have been put forward. Three of them involve the encounter of the Sun with one or more celestial objects. We will find that these theories have much of their structure based on assumption. It becomes rather clear that it is hard to think of a viable gravitational encounter mechanism to explain the Titius–Bode Law. Excluding the direct gravitational interaction theories mentioned at the end of this chapter, this emphasizes our viewpoint that if one believes in the validity of the Titius–Bode Law one probably must look for a nebular explanation of it.

A. Schmidt's Theory

In 1944 Otto Julievich Schmidt (1891–1956) proposed a new theory **(13.1)** in which the formation of the solar system was due to the Sun having captured a cloud of meteoritic material. [Later he modified his view to the position that the cloud could have been made of both dust and gas and might even have been a part of the original solar nebula **(13.2)**.]

For the building of the planets themselves, Schmidt felt that the captured meteorites slowly aggregated in competition with one another and was led thereby to propose an origin to the Titius–Bode Law **(13.3, 13.4)**.

Schmidt first defined the "boundary" position for an unaggregated meteorite as being the place where it has the same difference in angular momentum per unit mass with respect to the two proto-planets forming on either side of it. By this description, the angular momentum per unit mass of this boundary particle is given by

$$u'_n = \frac{u_n + u_{n+1}}{2},\qquad (13.1)$$

where u_n is the angular momentum per unit mass of the nth planet. Since, for any member of the solar system whose semi-major axis is a and eccentricity is e,

$$u = \sqrt{\{GMa(1-e^2)\}}, \tag{13.2}$$

a planet in a circular orbit has

$$u_n = \sqrt{(GMr_n)}. \tag{13.3}$$

Schmidt then recalled a result he obtained from his earlier accretion theory (13.1). This states that the distributions of the masses and orbits of the meteorites in an "angular momentum" region have the properties

$$dm = \frac{m}{2} de, \qquad -1 \leqslant e \leqslant 1, \tag{13.4}$$

$$a \equiv \frac{\varrho}{2} \left(\frac{1+e}{1-e} \right), \tag{13.5}$$

where ϱ is the limit distance at which capture occurred and is a constant for the system. Thus, by combining eqns. (13.4), (13.5), and (13.2), a meteorite has

$$u = \left(\frac{GM\varrho}{2} \right)^{1/2} (1+e), \tag{13.6}$$

or a boundary meteorite has (defining e'_n)

$$u'_n = \left(\frac{GM\varrho}{2} \right)^{1/2} (1+e'_n). \tag{13.7}$$

By equating the angular momentum of a planet to the total angular momentum of the constituent meteorites

$$u_n m_n = \int_{e'_{n-1}}^{e'_n} \left(\frac{GM\varrho}{2} \right)^{1/2} (1+e) \frac{m}{2} \, de$$

$$= \frac{1}{2} m_n [u'_{n-1} + u'_n], \tag{13.8}$$

77

where we have used

$$m_n = \frac{m}{2}(e'_n - e'_{n-1}) \qquad (13.9)$$

in obtaining the last line of eqn. (13.8). Combining eqns. (13.1) and (13.8) gives

$$u_n = (u_{n+1} + u_{n-1})/2 \qquad (13.10)$$

or

$$r_n^{1/2} = (r_{n+1}^{1/2} + r_{n-1}^{1/2})/2. \qquad (13.11)$$

The last line follows since $u_n \propto r_n^{1/2}$. Equation (13.11) can be recast in the form

$$r_n^{1/2} = \alpha + \beta n, \qquad (13.12)$$

which is Schmidt's Law, α and β being constants.

For the outer planets (from Jupiter on) Schmidt accepted this Law. His fit (in $AU^{1/2}$) to these five planets was good to 13% with $\alpha = 2 \cdot 28$ and $\beta = 1$. But for the inner planets Schmidt asserted that the Sun would capture most of the nearby material, either via gravitational collisions or because the material would lose angular momentum via the effects of radiation pressure, i.e. via the (John Henry, 1852–1914) Poynting–(Howard Percy, 1903–61) Robertson Effect (13.5). Schmidt claimed that the heavier asteroids that would be left would again form a Law like eqn. (13.12), but this time with $\alpha = 0 \cdot 62$ and $\beta = 0 \cdot 20$, fitting the four inner planets to 7%.

Of course there are many arbitrary points to this Law. Schmidt's Law comes from his choice of boundary condition, which in point of fact is similar in spirit to Kirkwood's spheres of attraction of a century earlier (6.4–6.7). But this boundary condition is at least questionable, in view of the work on the stability conditions of multiple-particle orbits that we have mentioned earlier (10.1, 10.7, 10.12). Next, Schmidt claims two Laws (why not a third for the intermediate region?), giving him four constants to work with, and still there was not tremendous agreement. We also have to believe that, although most of the mass of the inner cloud fell to the Sun, the distribution of the remainder was not perturbed enough to affect the forming of the same

Law on the inside. Finally, if the inner matter fell in and the outer matter did not, then the asteroid belts should be in a boundary region where we should expect a planet intermediate in size between the major and terrestrial planets. Certainly we should not get almost no material, as is the case.

Thus, for all these reasons and the fact that one must consider hydrodynamical processes in the gas–dust version, this theory cannot be accepted as an explanation of the Law.

B. Egyed's Theory

The theory of Egyed **(13.6)** can, if nothing else, be called the most unique proposal to explain the Titius–Bode Law. It is based on the old speculation of Dirac **(13.7)** that the gravitational constant may be changing with time. Dirac observed that

$$\frac{e^2}{Gm_p^2} = 1{\cdot}24 \times 10^{36} \approx O\left(\frac{m_e c^2 T}{\hbar}\right) = T_9(2{\cdot}42 \times 10^{37}), \quad (13.13)$$

where T (T_9 in 10^9 years = 1 eon) is the age of the universe as obtained from the (Edwin Powell, 1889–1953) Hubble constant.[†‡] Dirac felt that the approximate equality of the two large numbers in eqn. (13.13) was more than a coincidence and so suggested that G decreases inversely with time; i.e.

$$G(t) = GT/t. \quad (13.14)$$

Equation (13.14) means that earlier the Sun had a smaller radius (R), which Egyed assumed varies as

$$R = R_0 + \alpha t. \quad (13.15)$$

Equation (13.15) implies that the acceleration of gravity at the Sun's surface is

$$g_s(t) = \frac{GTM}{t(R_0 + \alpha t)^2}. \quad (13.16)$$

[†] At present this can be estimated **(13.8, 13.9)** to be about 10×10^9 years; i.e. $H = (94 \pm 9)$ km/(sec·Mpc) implies that $T = 1/H = (10{\cdot}4 \pm 1{\cdot}0) \times 10^9$ years.

[‡] Holmberg **(13.9)** gives a general discussion on methods of determining the Hubble constant.

However, the angular momentum $(\mathcal{L})$ is a constant,

$$\mathcal{L} \equiv \lambda = \dot{\theta} \sum_i m_i R_i^2. \tag{13.17}$$

Thus, the centrifugal acceleration at the solar equator varies as

$$a_c(t) = R\dot{\theta}^2 = \frac{R\lambda^2}{\left(\sum_i m_i R_i^2\right)^2} = \frac{A}{(R_0+\alpha t)^3}, \tag{13.18}$$

where A is a constant that depends on the particular mass distribution in the Sun. Comparing eqns. (13.16) and (13.18) we see that

$$X(t) \equiv \frac{a_c(t)}{g_s(t)} = \frac{A}{(R_0+\alpha t)} \frac{t}{GTM}. \tag{13.19}$$

For $(A - \alpha\, GTM) > 0$, there is a time t_1 when the centrifugal force becomes equal to and then greater than the gravity force $[X(t \geqslant t_1) \geqslant 1]$, so that part of the solar mass escapes. This causes the solar radius to decrease, which in turn makes g_s increase and a_c decrease. Thus, now $X\,(t > t_1) < 1$, so that the whole process can begin once more. Each loss is associated with a planet, the first being Pluto, the second Neptune, and so forth.

Egyed made the further assumption that

$$\Delta g_s(t_n) = K = \text{const.}, \qquad \Delta a_c(t_n) = K' = \text{const.}, \tag{13.20}$$

i.e. the increases of $g_s(t_n)$ and decreases of $a_c(t_n)$ are all respectively equal. Thus, after a number of mass losses (here it is ten) a_c can no longer equal g_s, so that no more planets will be formed. (A similar process is postulated for the satellite systems.)

Each planet will spiral out by conserving angular momentum as G decreases; i.e. since

$$(mr^2\dot{\theta})^2 = mr^2(mv^2) = mr^2\left(\frac{mMG(t)}{r}\right)$$

$$= m^2MGT\left(\frac{r(t)}{t}\right) = \text{const.}, \tag{13.21}$$

we have

$$r_n(t) = \frac{R(t_n)t}{t_n} = \left(\frac{R_0}{t}+\alpha\right)t, \tag{13.22}$$

where t_n is the time of the planet's mass being emitted. By then noting that a graph of g_s versus a_c could be made to have

$$t_n \simeq O(2t_{n-1}), \tag{13.23}$$

Egyed finally got

$$r_n = \left(\frac{R_0}{2^n t_0}+\alpha\right) T = \frac{C}{2^n} + D, \tag{13.24}$$

which is the Titius–Bode Law since the numbering is opposite to the usual.

Ingenious though the idea is, the theory cannot be accepted. This is firstly because of the great number of assumptions that are involved, assumptions that are explicitly geared to obtain the original Titius–Bode formulation. The assumptions of eqns. (13.15), (13.20), and (13.23) are what give the geometric progression, even though, for example, eqn. (13.23) is not even close to being exact. Similarly, the D term comes from the assumption of eqn. (13.15).

But even if all the assumptions could be accepted, detailed work that predates Egyed's paper rules out Dirac's speculation to begin with. The first argument is based on the work of Teller **(13.10)** who first discussed the effect that the resultant increased solar luminosity and smaller Earth orbital radius would have on the Earth's temperature. Using $T_9 = 2$, Teller showed that during the Cambrian era (1/2 eon ago) the Earth would have been at about 100°C so that there would not have been the life we know existed during the pre-Cambrian era. Even with the presently accepted age of the universe, the same argument **(13.11)** conflicts with the paleontological discovery **(13.12)** of bacteria and algae remains that date to 3·1 eons ago.

The second argument against a changing G constant comes from calculations on the evolution of the Sun **(13.13, 13.14)**. Pochoda and Schwarzschild **(13.13)** showed that even with the most favorable solar chemical composition, if the Sun is 4·5 eons old, it would have evolved into a red giant by now unless the age of the universe is greater than 15×10^9 years, conflicting with the Hubble constant age of the universe.

So, for all these reasons, Egyed's idea must be rejected.

Finally, if others are inclined to propose a similar theory, based on the possibility **(13.11)** that the electronic charge, from eqn. (13.13), varies with time as

$$e(t) = e\left(\frac{t}{T}\right)^{1/2},$$ (13.25)

we point out that one can reject this possibility for other strong experimental and theoretical reasons **(13.15, 13.16)**.

C. Woolfson's Theory

Soon after Egyed's paper appeared, Woolfson speculated **(13.17)** that the Titius–Bode Law was due to the encounter of a star with the Sun. He envisioned a star of 100 $M_\odot$ approaching to within ten solar radii of the Sun, with a relative velocity to it of 100 km/sec.

As Woolfson saw it, as the star approached the Sun would violate the Roche Limit. Thus, matter would be pulled out of the Sun which would eventually become Pluto. The tidal bulge on the far side of the Sun would then become a traveling wave. If it was a gravity wave on a deep liquid, it would take 80 minutes to reach a point nearest to the now closer star. Then a new extraction would take place, eventually to form Neptune. At the star's closest approach the matter of the asteroids would come out, most of it being captured by the star. As the star receded the matter of the inner planets would be pulled out. A similar mechanism would occur on the newly formed major planets, creating the satellite systems. Hence, the whole solar system would have been pulled from the Sun in 12 hours!

Woolfson put forward the above picture as a preliminary model, stating that a full consideration of it gives reasonable qualitative explanations for many of the physical properties of the solar system, including masses and orbital elements. He hoped the model would yield the Titius–Bode Law because of the regular time scale of mass emissions. But he deferred final judgement until after he had completed detailed mathematical calculations on the validity of the model.

However, when Woolfson's theory was finally published in mathematical detail (9.2, 13.18) the parameters and mechanism of the stellar encounter had been greatly changed. The incoming star now had a mass smaller than that of the Sun, and the planetary material was formed mainly from the incoming star, not the Sun. Further, there was no explicit mention of the encounter yielding a Titius–Bode Law.

Thus, as the author himself found that he had to reject the model as an explanation of the solar system and of the Titius–Bode Law, the idea must be abandoned.

D. Pendred's and Williams' Theory

Quite recently Pendred and Williams (13.19) offered a new theory of the solar system which is based on the slow accretion of matter from a diffused, rotating, dust–gas cloud that the Sun finds itself in just as it is created. In other words, the cloud is very large and partakes in the general galactic motion of the solar neighborhood. The gas feels the Sun's gravitational attraction and conserves angular momentum. However, radiation pressure and electromagnetic forces are ignored in treating the gas, as well as the attractive forces between molecules and grains. These are approximations which are open to criticism. The particular details of the condensation process in the Pendred–Williams theory are rather involved, but basically they are as follows:

In this theory the gas density is highest near the Sun, with a general decrease in density as the distance increases. Closer to the equatorial plane the density is higher and, in what is unique to this model, there is also a ring of much higher density in the equatorial plane at a distance $p_0(t)$, $p_0(t)$ moving outward with time.

This last phenomenon comes about thusly. Consider p_0 to be the

83

perihelion distance of an equatorial particle falling in from r'. Then

$$(2GMp_0)^{1/2} = \omega(r')^2, \tag{13.26}$$

where

$$\omega \gtrsim 10^{-15} \text{ rad/sec} = O[\omega(\text{Sun in galaxy})]. \tag{13.27}$$

For the limiting case of the particle being on a parabolic orbit, the time needed for the particle to reach perihelion from r' is

$$t = \tfrac{2}{3}(r')^{3/2} (2GM)^{-1/2}. \tag{13.28}$$

Combining eqns. (13.26) and (13.28),

$$p_0(t) = (\tfrac{9}{2})^{1/3} (\tfrac{9}{4}) \omega^2 (GM)^{1/3} t^{8/3}. \tag{13.29}$$

The idea is that all the gas which started on a sphere of radius r' falls in and reaches the equatorial plane near the distance $p_0(t)$ at approximately the same time, t. Thus, the gas produces a ring of maximum density at $r = p_0(t)$ and the ring moves out at a rate given by eqn. (13.29).

The gas ring is turbulent and therefore matter is condensed out of it at a high rate. This condensation is found to yield a mass distribution of

$$\frac{dm}{dp_0} = Ap_0^{-2}, \tag{13.30}$$

where A is a constant that depends both on the ratio of condensable to gaseous material and also on the exact form of the flow equation that is used.

Finally, the matter that condenses out of the ring when it is located at p_0 reaches equilibrium as grains at a distance $2p_0$. This is because the radial component of motion is zero when

$$
\begin{aligned}
0 &= r\dot{\theta}^2 - \frac{GM}{r^2} - \text{gas resistance} \\
&= \frac{GM}{r^2} \left(\frac{2p_0}{r} - 1\right) - \text{gas resistance},
\end{aligned}
\tag{13.31}
$$

so that for the condensed mass

$$r = 2p_0, \tag{13.32}$$

$$\frac{dm}{dr} = \frac{1}{2}\frac{dm}{dp_0}. \tag{13.33}$$

Pendred and Williams later used this model to try to explain the planetary masses and distances **(13.20)**. Integrating eqns. (13.30) and (13.33) gives the mass lying between $2p_1$ and $2p_2$ as

$$m(p_1, p_2) = \frac{A}{p_2}\left(\frac{p_2}{p_1} - 1\right). \tag{13.34}$$

If this all forms one planet, then conservation of angular momentum gives a final distance for the planet of

$$r(p_1, p_2) = \left[\frac{\int_{r_1=2p_1}^{r_2=2p_2} dr\left(\frac{dm}{dr}\right)(MG)^{1/2}r^{1/2}}{m(p_1, p_2)(MG)^{1/2}}\right]^2$$

$$= 8p_2\Bigg/\left[1 + \sqrt{\frac{p_2}{p_1}}\right]^2. \tag{13.35}$$

Pendred's and Williams' last step was to obtain a relationship between the p_2 of one planet and the p_1 of the next. They did this by defining what they called a "natural" gravitational reach, $R(p_1, p_2)$. This reach gives the distance out that a planet will attract grains, and therefore defines the boundary between one planetary region and the next. Despite some physical argumentation, it was admittedly arbitrarily defined as

$$2p_2 = r(p_1, p_2) + R(p_1, p_2) \tag{13.36}$$

$$\equiv r(p_1, p_2) + \mu[r(p_1, p_2)\, m(p_1, p_2)/M]^{1/3}, \tag{13.37}$$

μ being a constant and no symmetrical relationship for p_1 being defined. Equations (13.34)–(13.37) can be solved to give

$$p_2 = \mu\left(\frac{A}{M}\right)^{1/3}\frac{((p_2/p_1)-1)^{1/3}\left(1+\sqrt{(p_2/p_1)}\right)^{4/3}}{\left(1+\sqrt{(p_2/p_1)}\right)^2 - 4}, \tag{13.38}$$

where we have corrected the misprints that appeared in the originals of eqns. (13.36)–(13.38). Equation (13.38) is Pendred's and Williams' Law of Planetary Distances, since p_2 for one planet is p_1 of the next, and so on.

Pendred and Williams applied their theory to part of the solar system, taking $\mu/M^{1/3} = 0.25$, $A = 1/2$ for the inner planets, and $A = 1000$ for the outer planets. Their results are shown between the dashed lines of Table 13.1.

Leaving aside the questionable features of the theory that we have already mentioned, there are some very serious objections to this Law of Planetary Distances. In the first place, they use five constants [i.e. $\mu/M^{1/3}$, p_1(Venus), p_1(Jupiter), and the two A's] to fit six distances and still the results are off by up to 26%.

As stated, they only applied their theory to part of the solar system. When one considers the rest, worse troubles arise, as we have shown in Table 13.1. When applied to Mercury *their theory does not*

TABLE 13.1

The predictions of the Pendred–Williams theory.

Only the section between the dashed lines was published by them.
We have calculated and added the rest.

p_1	p_2	Mass	Distance	Planet	Observed mass	Observed distance
	0·182	(No solution)		Mercury	0·054	0·39
0·182	0·41	3·2	0·56	Venus	0·82	0·72
0·41	0·63	1·01	1·00	Earth	1·00	1·00
0·63	0·83	0·45	1·40	Mars	0·11	1·5
1·76	4·03	318	5·1	Jupiter	317	5·2
4·03	6·11	83	9·8	Saturn	95	9·5
6·1	7·95	38	14	Uranus	15	19
7·95	9·66	22·2	17·4	Neptune	17	30·1

even admit a solution. This is because

$$p_2(\text{Mercury}) \equiv 0 \cdot 182 < \lim_{(p_2/p_1) \to \infty} [\text{eqn. (13.38)}] = \mu \left(\frac{A}{M} \right)^{1/3} = 0 \cdot 265.$$

(13.39a)

For Neptune, which certainly is no minor particle in the solar system, *the distance is off by 42%*; and Pluto, the asteroids, and the satellite systems have yet to be dealt with.

So, even granting that the authors were also trying to calculate the planetary masses, they certainly have not yet explained the Law of Planetary Distances.

E. Hills' Theory

As we mentioned briefly in Chapter 10, Hills **(10.5)** attributes the Titius–Bode Law to a dynamical relaxation evolution of the planets from their original distances via point gravitational interactions.[†] Hills was motivated to accept this view by a number of observations.

He first noted that the Titius–Bode Law also seems to apply at least to the "regular" satellites. Since he rejected the view that the initial conditions were the same there as in the planetary system, he inferred that the Titius–Bode Law must have resulted from a period of strong gravitational encounters before the planets and satellites relaxed into stable orbits. In support of this view he pointed out that this is also indicated by the fact that many of the bodies in the solar system have large inclinations of their rotational angular momentum vectors with respect to their respective orbital angular momentum vectors.

To test this hypothesis Hills numerically studied the gravitational evolution of a number of planetary systems, each with a central star of mass $M_\odot$ and with arbitrary initial planetary orbits.

Hills made the following approximations (see Table 13.2) when he did his calculations. (1) The models were two-dimensional to simplify the numerical calculations. (2) The planet systems were all large-mass

[†] Hills' letter produced some amusing correspondences **(13.22, 13.23)**.

TABLE 13.2

The model planet systems that Hills studied and discussed.

Note that $M_{\text{Jupiter}} = 9 \cdot 54 \times 10^{-4} M$, and $M_{\text{Earth}} = 3 \cdot 039 \times 10^{-7} M$.

	Number of planets	Masses of the planets	Starting distances (AU)	Evolution time (yr)
"System 5"	4	all have $10\times$ the mass of a Jovian planet	$\approx 6, 9, 22, 29$	$4 \cdot 2 \times 10^3$
"System 9"	3	$8 \times 10^{-4} M$ $3 \times 10^{-4} M$ $1 \times 10^{-7} M$	$\approx 3, 5, 4$	$2 \cdot 5 \times 10^3$
"System 11"	6	all $5 \times 10^{-4} M$	All between $\approx (6-8)$	12×10^3

systems. This was to bring about a faster evolution. (3) The starting distances and eccentricities were chosen arbitrarily ("evolved from random initial orbits"). (4) Tidal effects were ignored. (5) No final stability was reached.

Thus, what Hills was really looking at were model, heavy, point, planet systems, hoping that the only difference this would make from the real solar system would be the length of time needed to obtain stability. In the models he studied he followed the evolution only over limited periods (see Table 13.2) and no final equilibrium was obtained. But after about 1 to 2×10^3 years there seemed to arise a tendency towards stability. This is demonstrated in Fig. 13.1 which shows the time variations of the semi-major axes of the planets in "System 11".

Hills then looked at the ratios of the periods of revolution of two pairs of adjacent planets. He plotted histograms of the number of times the period ratios were in a given interval, the evaluations taking place every 10 years. For these two pairs of planets he found that a large fraction of the time the ratios of the periods tended to be between 9/4 and 8/3, and they also had a second smaller tendency to be near the 3/2 commensurability (see Fig. 13.2).

88

Hills' interpretation was that this demonstrated a dynamical gravitational tendency towards commensurabilities, this being the entire origin of the Titius–Bode Law. If his interpretation is right, then, from eqn. (8.8), the (9/4 to 8/3) commensurabilities imply a progression ratio near (1·72 to 1·93) and the (3/2) progression commensurability implies a progression ratio near (1·145). One is also tempted to say this mechanism is the origin of not only our geometric progression ratio of 1·73 (of which Hills did not indicate an awareness), but also of the evolution function, since one might expect a complicated multistate set of "resonance relations" (see Chapter 8) to be set up.

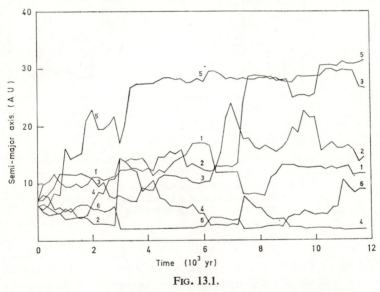

FIG. 13.1.

The time variation of the semi-major axes of the planets in Hills' "System 11".

Note further that the satellite systems have many more low-order commensurabilities than do the planetary and Hills' model systems. The satellite systems are low-mass systems where Hills expected this to happen, although he did not explain why.

89

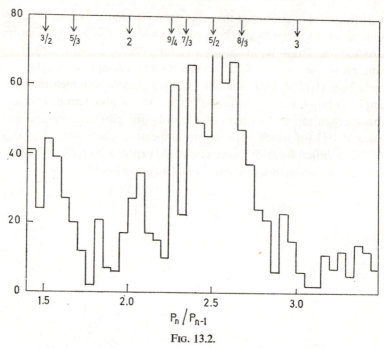

<figure>FIG. 13.2.</figure>

A histogram, in intervals of (1/20), of the ratio of the periods of the two outermost planets (P_n/P_{n-1}) in Hills' "System 11", except when the ratio is less than 1·4. In that case the third outermost planet is used in place of the second. The ratio is evaluated at intervals of 10 years, starting 1000 years after the beginning of the planetary evolution.

Assuming that no computational errors were made, Hills' results are very interesting and should be pursued. As such, there are a number of questions that have to be answered by proponents of such a theory.

1. What will be the results of a "realistic" calculation, i.e. with a model solar system of nine planets with the correct masses evolving in three dimensions? With the light planets and the third dimension, which often has unforeseen mathematical implications, the changes this would make could be significant.

2. Can this theory account for "gaps", i.e. the asteroid belts? In this

theory there is a connection from orbit to orbit via direct dynamical coupling. But remember from Fig. 7.1 that the Titius–Bode Law applies linearly down through to the low-mass planets. So, can this theory have evolved a gap through which the heavy outer and the light inner planets are still coupled by a Titius–Bode Law?

3. How does one justify this type of evolution *vis-à-vis* Poisson's Theorem (10.1–10.4) on the invariability of the semi-major axes? It is true that this Theorem applies only up to second order and for secular terms so that perhaps the short-range decay of the initial evolution does not have to be worried about. However, to turn things around, once stability is reached, if we were to reverse the orbits, going back to $t = 0$ would have to come from reinserting the instability whose origin we would still like to be explained.

Actually, at the time when Hills envisions the evolution, the planets would be at the end of Period II, the aggregation period. Thus, there might be some sort of the mixture of disks and large proto-planets we described earlier. Much of the gravitational relaxation might come from the evolution of such a complicated system, and so one could not concentrate on only the solid-planet evolution.

4. Even granting only solid-body evolution, since this evolution takes place at the end of Period II–beginning of Period III, when the planets would have been less compact and would have had larger atmospheres than they do now, would not tidal friction have played a large role in the evolution? That is to say, why were the planets not trapped in a tidal stability the first time they hit a reasonably low-order commensurability? We know, for example, that there exist long-distance tidal stabilities even among the small planets. Witness the coupling of the rotations of Mercury with the Sun and Venus with the orbit of Earth (13.21).

This being said, Hills has shown that gravitational evolution could have been important. However, if the evolution to the present distances was only from a geometric progression distance, the above objections would not be as difficult to overcome as if the evolution was from some random distance. It would be much easier for a gravitational mechanism to produce this smaller evolution. Thus, we still choose to view Hills' work as indicating that an evolution from geometric progression distances to the Blagg evolution function distances could have been possible via direct gravitational interactions, as well as via the tidal interactions discussed in Chapter 10.

However, one must be open to Hills' viewpoint which, of course, eventually could prove to be more correct. Further investigations are called for to help choose between the two outlooks and even to help to decide whether or not gravitational evolution alone could have produced the Titius–Bode Law. Such work might help solve one or both aspects of the Blagg–Richardson formulation.

In Chapter 15 we will suggest ways to attack this problem that might prove interesting and useful.

F. Dole's Theory

We finish this chapter with an account of a theory that was proposed by Dole (**13.24**). Due to a late printing, it became public as this book was in press. Since it involves a nebula as well as a gravitational driving mechanism, it could just as well be included at the end of Chapter 14. It is located here because of the similarity in spirit and execution to Hills' idea.

Dole considers a nebula containing both dust and gas that has a few percent of the solar mass and which is flattened into the shape of an exocone. The particles are considered to be in independent Keplerian orbits with random inclinations and orientations of the major axes, but all with the same assumed eccentricity, whose value is

$$e = (0.15 \text{ or } 0.25). \tag{13.39b}$$

The density distribution in the nebula is taken as (M = solar mass)

$$\varrho_2(\text{dust}) = A \exp(-\alpha r^\gamma), \tag{13.40}$$

$$A = (1.25 \text{ or } 1.50) \times 10^{-3}M, \tag{13.41}$$

$$\alpha = 5 \text{ (AU)}^{1/3}, \quad \gamma = 1/3, \tag{13.42}$$

$$\varrho_1(\text{gas})/\varrho_2(\text{dust}) = K-1, \tag{13.43}$$

$$K = (50 \text{ or } 100). \tag{13.44}$$

The mechanism for growth is the following: Planetary "nuclei" with masses $m_0 = 10^{-15}M$ are assumed to be injected one by one into the system. They have zero inclination to the ecliptic and prograde motion, but they have "random" semi-major axes and eccentricities. Their evolutions are followed by a computer, being governed by these rules:

The first nucleus, as it grows with mass $m \geqslant m_0$, is allowed to sweep up all the dust particles that come within a gravitational radius defined by

$$x = r[mM/(M+m)]^{1/4} \equiv r\mu^{1/4}. \tag{13.45}$$

In this way it sweeps out all of the dust within a certain region, making a dust-free ring in the nebula. Now, if by picking up the dust in an annulus the planet reaches a certain critical size (m_c), then it is allowed to capture gas. The critical mass is determined by a combination of (Sir James Hopwood, 1877–1946) Jeans's criterion for long term stability of gas in an atmosphere [see ref. **(13.25)**], assuming radiative equilibrium [see ahead to eqn. (14.12)] between the Sun and nebula and other simplifying physical assumptions. This leads to a critical mass given by

$$m_c = Br_p^{-3/4}, \tag{13.46}$$

$$B = (1 \cdot 2) \times 10^{-5} M \text{ (AU)}^{3/4}, \tag{13.47}$$

r_p and r_a being the perihelion and aphelion distances of the planet, respectively. Growth above the critical mass is determined by the effective density of the gas, which on physical grounds is taken to roughly be

$$\varrho = K\varrho_1 \bigg/ \left[(K-1)\left(\frac{m_c}{m}\right)^\beta + 1 \right], \tag{13.48}$$

$$\beta = \frac{1}{2}. \tag{13.49}$$

Boundaries of 0·3 AU and 50 AU are taken for the nebula.

A nucleus is injected and allowed to grow, either by just accumulating dust and not reaching the critical mass or by passing the critical mass and forming a large planet from gas. After one planet stops growing, another nucleus is allowed to start. This is continued sequentially until there is no dust left in the solar nebula, it being assumed the

^{r}emaining gas will evaporate. If, in the course of a later planet's evolution, it comes within a distance x of another existing planet, the two
planets are assumed to coalesce.

Dole did five sets of forty runs each. The simulated models yielded
smaller planets near the center of the solar system, larger planets
further out, and a few small planets near the edge. The results of selected
sets of runs are shown in Table 13.3.

TABLE 13.3

*Comparison of the solar system with three sets of Dole's computer-generated
systems.*

	Solar system	Set 3	Set 4	Set 5
Number of planets				
Mean	(9)	10·1	9·2	9·1
Range		7–12	7–11	7–11
Total mass of planets (in units of $10^{-3}M$)				
Geometric mean	(1·34)	1·16	1·56	1·71
Range		0·58–1·92	0·43–3·04	0·53–3·64
Mass of largest planet (in units of M_{Earth})				
Geometric mean	(317)	258	305	430
Range		90–594	63–979	89–1200
Spacing ratio				
Mean	"1·86"	1·73	1·84	1·86
Range	1·31–3·41	1·17–4·09	1·22–3·37	1·22–4·01

Dole argued that his results imply that the solar system is only one
out of a statistical set of possible planetary systems, his calculated
systems being other possibilities. But although there are a number of
tempting similarities between Dole's results and the solar system, they
still are open to a number of criticisms and open-ended questions:

1. In addition to the boundaries of the solar nebula, there are eight
other arbitrarily adjusted constants in the theory. To wit, K, A, α, γ, e,
m_0, B, and β.

2. The formulas in eqns. (13.40), (13.46), and (13.48) are *ad hoc*, even
though they are physically motivated.

94

3. The hydrodynamics and magnetohydrodynamics that play an extremely important part in the formation of the solar system are totally ignored (see Chapters 9, 12, and 14). In particular, such mechanisms are important in determining whether one can get large enough "nuclei" to create planets before the nebula evaporates. They also determine what the shape and mass of the nebula is in the first place.

4. The planet evolution is essentially two-dimensional. As noted in the last section, the predictions of an evolutionary process can be greatly affected when restricted to two dimensions. This is especially true since Dole leaves off where Hills takes up. It would be a fruitful examination to consider what Dole's primordial solar systems would do under further evolution. How would they evolve tidally and gravitationally? Further, how would they evolve if they were to try and create themselves not in isolated Keplerian orbits, but while there are interactions among the growing planets?

Dole felt that the solar system could be inconspicuously grouped with his results, but that is open to question. Looking at the solar system, one sees that the terrestrial planets go from small masses to larger masses back to small masses as the radius increases out to the asteroid belts. Then the largest giant planet follows, with the masses decreasing almost monotonically out to Pluto. None of Dole's published individual runs have these properties.

Finally, one can show how Dole's mechanism would automatically imply smaller planets near the Sun. Ignoring the exponential fall off in density to make the argument simpler, one can see, by comparing

$$m_c \cong \text{const.} \, (r_a^2 - r_p^2) \equiv \text{const.} \, r_p^2 (\Delta r)_c \qquad (13.50)$$

with eqn. (13.46), that

$$(\Delta r)_c = \text{const.}/r_p^{11/4}. \qquad (13.51)$$

Equation (13.51) shows that as r gets larger, the Δr needed to obtain the critical mass is smaller. In other words, the further out one is, the smaller is the necessary eccentricity for the planet in order to allow the critical mass to be reached. Further, the change in energies of particles that need to be captured is smaller. One sees this by noting that

$$\Delta E \equiv E_a - E_p \cong \text{const.} \, (\Delta r)/r_p^2. \qquad (13.52)$$

95

Therefore, it is physically clear that one has to reach a certain distance before large planets can be made in this model, and that distance is determined by the input constants of the system.

With the uncertainties we have mentioned, Dole's theory cannot be said to have solved the problem. However, his work does give us an interesting model to contemplate and it could prove to be at least partially headed in the right direction. Note that it claims that there is no real order in the original solar system, the distances being only partially ordered by the two accretion mechanisms. If this is true, one would have to give up a "nebular" origin to the Titius–Bode Law, although the later gravitational and tidal evolutions would remain an open question.

CHAPTER 14

Nebular Theories

A. Berlage's Nebular Theory

In Chapter 12 we discussed the electromagnetic theories of Berlage. As mentioned, he eventually abandoned that line of attack since **(12.5)** he could not explain the satellite systems and there were also other difficulties in making the mechanism work.

However, in the meantime, Berlage had already **(14.1)** begun to think about a nebular origin to the Titius–Bode Law. Once again his theories anticipated later proposals and were not recognized as having done such.

Berlage started by considering a nebula rotating about the z-axis going through the Sun, and under the influence of gravitation and internal gas pressure. The equation of equilibrium of this system is

$$\nabla\left(\frac{GM}{\mathcal{R}}\right) + \frac{1}{\varrho}\nabla P = \omega^2\mathbf{r}, \qquad (14.1)$$

$$\mathcal{R}^2 \equiv r^2 + z^2, \qquad (14.2)$$

or writing the components of eqn. (14.1) separately,

$$\frac{GMz}{(r^2+z^2)^{3/2}} + \frac{1}{\varrho}\frac{\partial P}{\partial z} = 0, \qquad (14.3)$$

$$\frac{GMr}{(r^2+z^2)^{3/2}} + \frac{1}{\varrho}\frac{\partial P}{\partial r} = \omega^2 r. \qquad (14.4)$$

Introducing the Perfect Gas Law (μ being the chemical weight)

$$P = \frac{\varrho kT}{\mu m_p} \equiv \mathcal{K}\varrho T \qquad (14.5)$$

into eqn. (14.3) and integrating it with respect to z gives

$$\ln P = \ln P_e - \frac{GM}{\mathcal{K}}\int_0^z \frac{h\,dh}{T(r^2+h^2)^{3/2}}, \qquad (14.6)$$

97

where the subscript e refers to points on the equatorial plane, i.e.

$$P_e(r) = P(r, z = 0). \tag{14.7}$$

By now differentiating eqn. (14.6) with respect to r and substituting it into eqn. (14.4) one obtains

$$\frac{T}{T_e}\frac{GM}{r^2} + \mathcal{K}T\frac{d \ln P_e}{dr} + GMT \int_0^z \frac{\left(h\frac{\partial T}{\partial r} - r\frac{\partial T}{\partial h}\right)}{T^2(r^2+h^2)^{3/2}}\, dh = \omega^2 r. \tag{14.8}$$

A general simplification of eqn. (14.8) is the situation where the temperature is a function only of $\mathcal{R}$. Then the numerator in the integral of eqn. (14.8) vanishes so that

$$\frac{T}{T_e}\frac{GM}{r^2} + \mathcal{K}T\frac{d \ln P_e}{dr} = \omega^2 r. \tag{14.9}$$

For a specific example, Berlage first considered the case of isothermy ($T = T_e$ = constant). Using eqn. (14.5) to eliminate P_e for ϱ_e,

$$\frac{GM}{r^2} + \frac{\mathcal{K}T}{\varrho_e}\frac{d\varrho_e}{dr} = \omega^2 r. \tag{14.10}$$

In this isothermal case eqn. (14.10) shows that the nebula is freely rotating, i.e. $\omega = \omega(r)$ is a function only of the distance from the axis. Further, from eqns. (14.5) and (14.6) we then get that both P and ϱ are of the form

$$\begin{Bmatrix} P \\ \varrho \end{Bmatrix} = \begin{Bmatrix} P_e \\ \varrho_e \end{Bmatrix} \exp\left[-\frac{GM}{\mathcal{K}T}\left(\frac{1}{r} - \frac{1}{\mathcal{R}}\right)\right]. \tag{14.11}$$

This solution is a thin disk since, for example, air at $23 \cdot 2°K$ would have $\varrho(r, 0\cdot 1r) \cong 10^{-10}\varrho_e(r)$ at the distance of Neptune.

Later **(14.2)** Berlage came to the conclusion that T = constant was inconsistent with steady motion. So, he accepted the more realistic case of radiative equilibrium where (Γ = const.)

$$T = \Gamma \mathcal{R}^{-1/2}. \tag{14.12}$$

Eqn. (14.12) results because the radiation absorbed is proportional to $\mathcal{R}^2$ and the radiation emitted is proportional to T^4. Since he could have steady motion and keep the same eqn. (14.10) if μ was taken to vary as

$$\frac{T}{\mu} = \text{const.,} \qquad (14.13)$$

Berlage took eqn. (14.13) as his working hypothesis.

To get ϱ_e, Berlage looked at the mass distribution of the planets and took as a first guess

$$\varrho_e = \varrho_0 e^{-ar}. \qquad (14.14)$$

However, he was able to improve on this guess by investigating the consequences of internal friction in the gas. If nothing else, the friction is due to the differences in Keplerian angular velocity as a function of distance. The idea is the following:

First, the rate of change of angular momentum due to the viscous transfer of molecules between two rings is taken to be zero, i.e.

$$\frac{d}{dt}(L_\eta) = \frac{d}{dr}\left(\eta r^3 \frac{d\omega}{dr}\right) dr\, d\theta\, dz = 0, \qquad (14.15a)$$

$$\left(\eta r^3 \frac{d\omega}{dr}\right) = \text{const.,} \qquad (14.15b)$$

where η is the coefficient of viscosity. Then this is combined with the condition of a minimum loss of energy due to the friction **(14.3)**. For the cylindrical symmetry involved here, this condition is

$$\delta\left(\frac{dE}{dt}\right) = \delta \iiint \eta \left(\frac{d\omega}{dr}\right)^2 r^3\, dr\, d\theta\, dz. \qquad (14.16)$$

If one takes an indefinite volume bounded by two coaxial cylinders and two plates parallel to the ecliptic, eqn. (14.16) implies

$$\delta\left[\eta\left(\frac{d\omega}{dr}\right)^2 r^3\right] = 0 \qquad (14.17)$$

or, combining eqns. (14.15) and (14.17),

$$\delta\left(\frac{d\omega}{dr}\right) = 0. \qquad (14.18)$$

A solution can now be obtained for ϱ_e by combining eqn. (14.18) with eqn. (14.10) and varying ω slightly from

$$\omega_1 = \left(\frac{GM}{r^3}\right)^{1/2} \tag{14.19}$$

to

$$\omega_2 = \left[\frac{GM}{r^3} + \frac{\mathcal{K}T}{r}\frac{d\ln\varrho_e}{dr}\right]^{1/2}. \tag{14.20}$$

Then, substituting eqns. (14.19) and (14.20) into eqn. (14.18) gives

$$\frac{1}{\omega_2}\left(\frac{-3GM}{r^4} - \frac{\mathcal{K}T}{r^2}\frac{d\ln\varrho_e}{dr} + \frac{\mathcal{K}T}{r}\frac{d^2\ln\varrho_e}{dr^2}\right) = \frac{1}{\omega_1}\left(\frac{-3GM}{r^4}\right). \tag{14.21}$$

To first order in $\left(\dfrac{\mathcal{K}T}{\omega_1^2 r}\dfrac{d\ln\varrho_e}{dr}\right)$ this is

$$\frac{d\ln\varrho_e}{dr} + 2r\frac{d^2\ln\varrho_e}{dr^2} = 0, \tag{14.22}$$

yielding the solution (ϱ_0 and a are constants)

$$\varrho_e = \varrho_0 e^{-ar^{1/2}}. \tag{14.23}$$

Berlage's initial proposal for this yielding a Titius–Bode type Law came from considering a small ($\varepsilon \ll 1$) oscillatory modification to eqn. (14.23), i.e.

$$\varrho_e = \varrho_0 \exp\left[-ar^{1/2} + \varepsilon \sin f(r)\right], \tag{14.24}$$

and demanding that when eqn. (14.24) was substituted into eqn. (14.22) the result would remain small (i.e. power-wise independent of r). This gave (b and c are constants)

$$f(r) = br^{1/2} + c, \tag{14.25}$$

or, by having the planets correspond to the positions of maximum of $f(r)$,

$$\sqrt{r_{n+1}} - \sqrt{r_n} = \frac{2\pi}{b}, \tag{14.26}$$

which is the same as Schmidt's Law of eqn. (13.12).

100

However, later **(14.4)** Berlage realized that he really needed a power-independent result when eqn. (14.24) was substituted into eqn. (14.22) multiplied by r. Only then is the resulting equation unitless. This gives

$$f(r) = b \ln r + c, \tag{14.27}$$

with the result that

$$\ln r_{n+1} - \ln r_n = \frac{2\pi}{b},$$

$$\left(\frac{r_{n+1}}{r_n}\right) = \exp\left(\frac{2\pi}{b}\right), \tag{14.28}$$

i.e. the classical Titius–Bode Law.

To summarize, Berlage's final theory was that the dynamical equation (14.1) combined with the T and μ eqns. (14.12) and (14.13) yields a rotating spherical nebula concentrated mainly in a thin disk with a density given by eqn. (14.11). When combined with the restriction of equilibrium, i.e. no transfer of angular momentum by friction and a minimum energy loss by friction, this yields the density function for ϱ_e given in eqn. (14.23). Forced deviations from eqn. (14.23) will be small if they are of the oscillatory character given by eqns. (14.24) and (14.27). The interpretation is that preferential condensation will occur near the positions of maxima of $\sin[f(r)]$, due to the higher density, and so that is where the planetary primordial rings will be. From eqn. (14.27) this yields the Titius–Bode Law.

The theory does propose an interesting mechanism, but firstly it really should have considered the case of

$$T = \Gamma \mathcal{R}^{-1/2}, \qquad \mu = k/(\mathcal{K}m_p) = \text{const.} \tag{14.29}$$

This is a better model of the original nebula. In this case, as **Berlage** mentioned **(14.4)**, the equations that result are

$$\varrho = \varrho_e \left(\frac{\mathcal{R}}{r}\right)^{1/2} \exp\left[-\frac{2GM}{\mathcal{K}\Gamma}\left\{\frac{1}{r^{1/2}} - \frac{1}{\mathcal{R}^{1/2}}\right\}\right], \tag{14.30}$$

$$\frac{GM}{r^2} + \frac{\mathcal{K}\Gamma}{r^{1/2}}\frac{d\ln\varrho_e}{dr} - \frac{\mathcal{K}\Gamma}{2r^{3/2}} = \left(\frac{\mathcal{R}}{r}\right)^{1/2}\omega^2 r. \tag{14.31}$$

If for $z = 0$ these equations are handled as were eqns. (14.14) to (14.23), then the minimum energy loss method gives

$$r^2 \frac{d^2 \ln \varrho_e}{dr^2} + \frac{1}{2} = 0, \tag{14.32}$$

or

$$\varrho_e = \varrho_0 \left(\frac{r}{r_0} \right)^{1/2} e^{-ar}. \tag{14.33}$$

But if now an oscillatory function is put in the exponential of eqn. (14.33), then the solution for $f(r)$ given in eqn. (14.27) will still yield a small result, power-wise independent of r, when substituted into eqn. (14.32). Thus, the Titius–Bode Law of eqn. (14.28) still holds.

However, even with this improvement, the theory rests on two important assumptions. The first is that the deviations of ϱ_e from equilibrium will be of the form exp [sin $f(r)$] instead of other, possibly complicated, terms that include strong transients. It is this assumption that yields the Titius–Bode Law. Further, there is the assumption that this deviation, once given, will be enough to cause preferential condensation into Titius–Bode rings, even though by definition the amplitude of this deviation is small. Here the theory lacks quantitative argumentation, especially since Berlage envisioned a stable system. The reason is that turbulence is extremely important in the early stages of the nebula and condensation is favored in turbulent regions.

Finally, even granting that one can ignore other processes, such as electromagnetic ones, the values of the normalization constant and the geometric progression constant of the Titius–Bode Law are not explained dynamically. This, of course, is necessary to really derive the Law.

In later years, Berlage's interest in the subject remained (14.5–14.7). He concentrated his attention on the mass distributions, groupings, and distances of the planets and satellites and also focused on nebular theories such as his own and those of the turbulence variety. Eventually he became convinced of the importance of turbulence in the early stages of the nebula and also that most of the matter was lost by evaporation **(14.6)**. So, much of his later discussion **(14.6, 14.7)** followed the ideas of the turbulent theories we come to next. However, he still felt that his early ideas were applicable to the material that remained after evaporation (i.e. during our early Period II) and that the answer to the Titius–Bode Law was here, since he could not see how turbulence could produce such a regular Law.

To sum up, although this theory can be regarded as interesting, it must of necessity be regarded as a model that certainly, at least, does not apply to the early nebula period. However, it was an indicative precursor to the turbulent theories and could possibly apply to the late stages of Period I and the early stage of the aggregation Period II if enough of the disk was in the form of a non-turbulent gas.

B. The von Weizsäcker Theory

In the closing stages of World War II, C. F. von Weizsäcker published a new nebular theory on the origin of the solar system **(14.8)** which caused much excitement. It was later described elsewhere **(14.9, 14.10)** to allow non-German scientists to become acquainted with it.

Significantly, von Weizsäcker's theory was the first to attribute to the Sun and its nebula the type of chemical composition that we now know to be correct[†] **(14.11)**. At that time most people thought that there was a vanishing helium content in the Sun and that about 65% of its weight was due to heavy elements. However, earlier Strömgren **(14.12)** had pointed out that a possible solar composition was 55% H, 40% He, and 5% higher elements by weight. A paper by Biermann **(14.13)**, indicating that 99% of the Sun's mass was made up of hydrogen, helium, and possibly oxygen, vindicated this view and convinced von Weizsäcker that he should consider such a system.

For there to have been enough heavy elements in such a nebula to

[†] Ref. **(14.11)**, which is a current standard on the subject, gives the percentage compositions by weight in the Sun as H $= 74.4\%$, He $= 24.07\%$, C $= 0.5\%$, and O $= 1.1\%$.

have made the planets, von Weizsäcker felt its mass had to have been about 10% of the solar mass, and further that later most of it evaporated. von Weizsäcker's theory was the first to take this position, and other theories, such as Hoyle's **(9.4)**, are successors to it. However, von Weizsäcker did not attempt to explain the nebula's formation but assumed its existence and then asked what its properties would be.

To do this he considered the equilibrium equations (14.1)–(14.4) under the eqn. (14.29) assumptions of radiative equilibrium and uniform chemical composition. Then the cylindrical state equations become

$$\mathcal{K}\Gamma \frac{\partial \ln \varrho}{\partial r} = r\mathcal{R}^{1/2}\left(-\frac{GM}{\mathcal{R}^3} + \frac{\mathcal{K}\Gamma}{2\mathcal{R}^{5/2}} + \omega^2\right), \qquad (14.34)$$

$$\mathcal{K}\Gamma \frac{\partial \ln \varrho}{\partial z} = z\mathcal{R}^{1/2}\left(-\frac{GM}{\mathcal{R}^3} + \frac{\mathcal{K}\Gamma}{2\mathcal{R}^{5/2}}\right). \qquad (14.35)$$

On changing the variable in eqn. (14.35) from z to $\mathcal{R}$ we have

$$\mathcal{K}\Gamma \frac{\partial \ln \varrho}{\partial \mathcal{R}} = -\frac{GM}{\mathcal{R}^{3/2}} + \frac{\mathcal{K}\Gamma}{2\mathcal{R}}, \qquad (14.36)$$

which can be integrated to give [$\tau(r)$ being the integration constant]

$$\mathcal{K}\Gamma \ln \varrho = \frac{2GM}{\mathcal{R}^{1/2}} + \frac{1}{2}\mathcal{K}\Gamma \ln \mathcal{R} + \tau(r). \qquad (14.37)$$

If we put this into eqn. (14.34) we have

$$\frac{\partial \tau(r)}{\partial r} = r\mathcal{R}^{1/2}\omega^2, \qquad (14.38)$$

so that

$$\omega^2 = \left[\frac{\partial \tau(r)}{\partial r}\right]r^{-1}\mathcal{R}^{-1/2} \equiv W(r)\mathcal{R}^{1/2}. \qquad (14.39)$$

Thus, ω^2 is given by a function of r multiplied by $\mathcal{R}^{-1/2}$. The solution that von Weizsäcker considered, and note that this is an assumed solution, is

$$\omega^2 \equiv GMq^2(r)r^{-5/2}\mathcal{R}^{-1/2}. \qquad (14.40)$$

104

$q(r)$ remains to be determined, but when $q \equiv 1$, eqn. (14.40) exactly describes a Kepler orbit due to the Sun's gravitational attraction in the plane of the ecliptic. By now putting eqn. (14.40) into eqn. (14.39), integrating, and substituting into eqn. (14.37), the form of $\varrho(r)$ is found to be

$$\mathcal{K}\Gamma \ln \varrho = -2GM\left(\frac{1}{r^{1/2}} - \frac{1}{\mathcal{R}^{1/2}}\right) + \frac{1}{2}\mathcal{K}\Gamma \ln \mathcal{R} + \text{const.}, \quad (14.41)$$

$$\varrho = \varrho_0\left(\frac{\mathcal{R}}{r_{\max}}\right)^{1/2} \exp\left[-\frac{2GM}{\mathcal{K}\Gamma}\left(\frac{1}{r^{1/2}} - \frac{1}{\mathcal{R}^{1/2}}\right)\right], \quad (14.42)$$

where $\varrho_0 \times (r_{\max})^{-1/2}$ is a normalization constant. [Note that eqn. (14.42) agrees with the "minimum energy" solution of eqns. (14.30) and (14.33) in the case where $a = 0$.]

If q really were equal to unity over all space, then the density in the plane of the ecliptic would become infinite at infinity. To correct this, von Weizsäcker took

$$q(r) = \begin{cases} 1 & r < r_{\max} \\ 0 & r > r_{\max}. \end{cases} \quad (14.43)$$

Then, by continuity of the solution of eqn. (14.42) from integrating eqn. (14.39), this gives

$$\varrho = \varrho_0\left(\frac{\mathcal{R}}{r_{\max}}\right)^{1/2} \exp\left[-\frac{2GM}{\mathcal{K}\Gamma}\left(\frac{1}{r^{1/2}} - \frac{1}{\mathcal{R}^{1/2}}\right)\right] \quad r < r_{\max}, \quad (14.44a)$$

$$\varrho = \varrho_0\left(\frac{\mathcal{R}}{r_{\max}}\right)^{1/2} \exp\left[-\frac{2GM}{\mathcal{K}\Gamma}\left(\frac{1}{r_{\max}^{1/2}} - \frac{1}{\mathcal{R}^{1/2}}\right)\right] \quad r > r_{\max}. \quad (14.44b)$$

Thus, the picture is that for $r < r_{\max}$ there is a rotating disk. The pressure ($P = k\Gamma\varrho\mathcal{R}^{-1/2}$) is constant on the equator but falls off exponentially at right angles to it. For $r > r_{\max}$ the pressure is spherically symmetric and falls off exponentially in a barometric manner. The normalizations that von Weizsäcker used were

$$r_{\max} = 4 \times 10^{14} \text{ cm},$$
$$\varrho_0 = 10^{-9} \text{ g/cm}^3, \quad (14.45)$$
$$\Gamma = (300°\text{K})(10^{13} \text{ cm} = \text{distance to Venus})^{1/2},$$

$$\mu = \frac{k}{\mathcal{K}m_p} \simeq 4.$$

105

It can be seen that the disk is thin under these circumstances by noting that at planetary distances eqn. (14.44a) has the density falling off by ϱ^{-2} for

$$\frac{z}{r} \simeq r^{1/4}\left(\frac{2\mathcal{K}I}{GM}\right)^{1/2} \simeq \frac{1}{30}. \tag{14.46}$$

von Weizsäcker then studied the hydrodynamical properties of the nebula and came to the conclusion that viscous forces due to differential rotation caused turbulence and a resulting loss of energy in the system. This implied that the nebula lost mass both to infinity and onto the Sun. The lifetime of the nebula under the effects of gravity and viscous forces was found to be about 10^7 years. The reader is referred elsewhere for a further discussion (**14.8–14.10, 14.14**) but should note that it is during this nebular period of 10^7 years (our Period I) that von Weizsäcker's theory had to produce the Titius–Bode Law.

von Weizsäcker obtained the Titius–Bode Law from his theory by asking the following question: Is it possible to have a nebula that has a minimum energy loss from viscosity in such a way that it also roughly obeys Keplerian motion within certain regions, hence producing a state of "quasi-stability"? von Weizsäcker was able to construct one situation that gave a positive answer and yielded the Titius–Bode Law.

Consider two gas particles with the same energies [or equivalently, to the first order, the same constants of area (**14.10**)], but one being in a circular orbit of radius r_0 and the second having a slightly eccentric orbit. Let $t = 0$ be when the two particles are along the same radius vector from the Sun and separated by the maximum distance Δ. If one transfers to a rotating coordinate system centered on the particle in a circular orbit, it is well known (**14.15**) that with an inverse square field the second particle describes an ellipse about the first given by

$$X = \Delta \cos \omega t,$$
$$Y = -2\Delta \sin \omega t, \tag{14.47}$$
$$\omega = \left(\frac{GM}{r_0}\right)^{1/2}$$

106

or

$$X^2 + \tfrac{1}{4}Y^2 = \Delta^2. \tag{14.48}$$

Thus, the ellipse has a major axis in the transverse direction with a length 4Δ that is twice the length of the radial axis, and the motion is in the opposite sense to the orbital rotation (see Fig. 14.1).

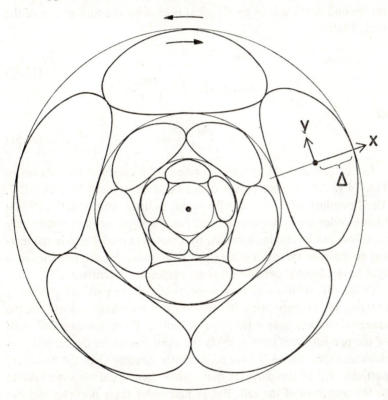

FIG. 14.1.

The type of system of regular vortices that von Weizsäcker's theory predicts. Each of the closed curves represents the shape of a vortex cell in the inertial coordinate system, or the shape of the orbit of a particle bounding the cell in its particular rotating coordinate system (the coordinates label a boundary particle in such a system). Note that we use von Weizsäcker's figure for five vortices in a ring, although the text shows that six is a more acceptable number.

The eccentricity (e) of the orbit can be related to Δ by using the equation of an ellipse with semi-major axis a,

$$r = \frac{ea}{(1 + e\cos\theta)}, \qquad (14.49)$$

and remembering that in the solar system's (inertial) coordinate system the second particle is in an elliptical orbit with the Sun as one of the foci. Thus,

$$r_{max} = r_0 + \Delta = \frac{ea}{1-e},$$
$$r_{min} = r_0 - \Delta = \frac{ea}{1+e}, \qquad (14.50)$$

or

$$\Delta = \frac{e^2 a}{1 - e^2} = e r_0. \qquad (14.51)$$

Let us consider the gas within one of the closed curves shown in Fig. 14.1. All of these curves define what we will call "vortex cells". The boundary of a vortex cell roughly satisfies eqn. (14.48). (When higher-order terms are considered, the shape of a cell is not exactly an ellipse but it remains bounded by a closed curve.) Thus, in the inertial coordinate system a vortex cell is bounded by the particles in it that have elliptical orbits with some maximum eccentricity.

Originally, within a given volume, there will be particles with many energies and eccentricities. But friction will eventually smooth out the energy distribution as a function of radius. Then, within a cell most of the particles will have roughly the same energy as the central particles and they also will have eccentricities smaller than the boundary particles. All of the particles that satisfy these conditions we take to be the members of the cell. But to first order these particles will not interfere with each other's orbit and so are in free Keplerian motion relative to one another. Thus, this is a picture where appreciable viscous effects exist only at the inner and outer boundaries of the vortex cell. (One can think of an analogy to whirls in a flowing river.)

When a vortex is first made, it is small, and thus it has only particles with small eccentricities in it. So, it will most often encounter particles

in orbits of roughly the same energy but with larger eccentricities•
But these particles can be easily captured by the vortex, since their
orbits will be outside the cell and so can be attached to the cell without
disrupting it. (It is easiest to see this in the rotating coordinate system,
where the new particles will describe ellipses surrounding the cell,
forming, so to speak, a new layer of skin for the cell.) Eventually,
however, as the vortex grows, it will more often encounter particles
with different energies and smaller eccentricities whose capture would
disrupt the cell, and so the vortex's growth is limited.

What von Weizsäcker pictures is a series of rings, each bounded
by r_{n+1} and r_n, in which a definite number of vortex cells exist. It is at
the r_n where the greatest turbulence occurs (the contact point of adjacent rings and hence vortices) and where the condensation of the even-
tual planetary material would occur. Further, since r_{n+1} and r_n bound
a vortex cell, they are on the orbit of maximum eccentricity within
a vortex. So, from eqn. (14.50)

$$\frac{r_{n+1}}{r_n} = \frac{1+e_{max}}{1-e_{max}}. \tag{14.52}$$

Equation (14.52) shows that $e_{max} = 1/3$ corresponds to the classical
Titius–Bode progression of 2. von Weizsäcker liked this since
$e_{max} = 1/3$ would mean that the differences between the velocities of
the vortices in adjacent rings at a boundary is

$$\Delta v = v_{outer} - v_{inner} = \left(\frac{GM}{r_n}\right)^{1/2} [\sqrt{(1+e_{max})} - \sqrt{(1-e_{max})}]. \tag{14.53}$$

This is 10 km/sec at the distance of Earth and 3 km/sec at Jupiter,
so that turbulence would be expected. Further, since $\Delta v > 0$, the
secondary eddies would be in the contrary sense to the primary eddies
and so in the same sense as the planetary orbits. Thus, these secondary
eddies, which would form the planets, should give the planets prograde
spins.

The vortices, once formed, need to be stable for a period of about
10 to 100 years. If they are, then particles that are large enough not to
be swept along by turbulence after the cells decay can be formed at

109

the interfaces of the rings. This would allow the formation of the planets.

The number of vortices in a ring (N) can be found from eqn. (14.51). It is given by the relation

$$e_{max} = \frac{\Delta}{r_0} \simeq \sin\left(\frac{1}{4} \text{ angular size of a cell}\right) = \sin\left(\frac{\pi}{2N}\right), \quad (14.54)$$

which is good to 1% for $e_{max} \leqslant 1/3$. In Table 14.1 we have given the number of regular vortices in a ring and the geometric progression ratio for various e_{max} in the von Weizsäcker theory.

TABLE 14.1

The number of vortices in a ring (N) and the geometric progression ratio implied in the von Weizsäcker theory for various e_{max}.

e_{max}	N	r_{n+1}/r_n
0·383	4	2·24
1/3	4·63	2·00
0·309	5	1·89
0·259	6	1·70
0·220	7	1·57
0·195	8	1·49

von Weizsäcker noted that it was not necessary that all systems have the same number of vortices in a ring. The numbers could be different, thus yielding different progression ratios. As examples, the satellite systems seemed to him to be places where a lower progression ratio applied, thus implying that there had been more vortices in a ring. The number of vortices would be determined by the detailed dynamics of the particular nebula, but the basic formation process, and hence the making of the Titius–Bode type Law, would be the same.

Be this as it may, in von Weizsäcker's original paper, and in the later discussions of his theory, comment has always centered on the system of five regular vortices in a ring. This is because the closest that a progression ratio from a system of regular vortices can come to 2 (the original Titius–Bode proposal) is 1·89 with five regular vortices. However, from Table 14.1 we see that a regular system of six vortices gives a progression ratio of 1·70. This is very close to the Blagg–Richardson number of 1·73, which is the best fit to the planets and the Jupiter system.

We have here an example of how too much of the discussion of the Titius–Bode Law has been tied to the classical progression ratio of 2, when it need not be. At least for the planets, a regular system of six vortices is clearly preferable to five vortices. Having six vortices does not detract from von Weizsäcker's theory, but actually enhances it.

von Weizsäcker's theory is attractive since it offers an explanation for the geometrical progression in terms of the type of chemical and dynamically turbulent solar nebula that we now feel is correct. Further, the mechanism in itself is not *ad hoc*. It comes from trying to minimize the energy loss in a realistic, turbulent nebula in such a way as to allow the solar system to be formed.

The whole theory, however, hinges on one intuitive physical assumption and on one argument that is open to debate. The assumption is that a regular system of vortices will be set up to minimize the energy loss instead of having some other more complicated system of vortices being set up. The argument is that the vortices will last 10 to 100 years. This is needed so that after the vortices decay, the distribution of the condensed matter will not be destroyed by the ensuing disruptive turbulence (the sizes of the condensed particles will have become large enough to resist the turbulence).

In the decade following the appearance of von Weizsäcker's theory, discussion of it eventually concentrated on trying to answer the questions raised by these two points. This work produced sophistications of the original theory and also a counter theory, which we shall discuss in turn.

C. Modifications of the von Weizsäcker Theory

After World War II attempts were made to put von Weizsäcker's explanation of the Titius–Bode Law on a better footing by considering the theory of turbulence[†] **(14.16)**. The first suggestion was made by Tuominen **(14.17)**, who showed that both the lifetimes and the sizes of the eddies that existed in the solar nebula were of the orders of magnitude desired in the von Weizsäcker theory.

First, he pointed out that the lifetime (τ) of a vortex cell can be roughly described by (Ludwig, 1875–1953) Prandtl's formula **(14.18)**

$$\tau = \frac{l}{u} = \frac{1}{\left| \dfrac{dv}{dr} \right|}, \tag{14.55}$$

where v describes the velocity of the stream lines, and a turbulent mass element moves perpendicular to the stream lines with a velocity u in the direction r over a distance l before it loses its identity (i.e. l is the mixing length). If we take v to be given by the Kepler velocity

$$v_K = \left(\frac{GM}{r} \right)^{1/2}, \tag{14.56}$$

then

$$\tau = \frac{2r^{3/2}}{(GM)^{1/2}} = \tau_K/\pi, \tag{14.57}$$

where τ_K is the Kepler period of revolution.

This result, of course, follows from how eqn. (14.55) is interpreted. But in an elementary manner it still shows that one could have stable vortex cells for at least on the order of years, at the distances of the major planets. Note further that u and l are considered as linear quantities, whereas in von Weizsäcker's theory they are modified by the shape of the closed curves of eqn. (14.48) or Fig. 14.1.

Tuominen also argued that the sizes of the eddies could be esti-

[†] For an excellent discussion on the theory of turbulence with a large bibliography the reader is referred to ref. **(14.16)**.

mated by means of (Theodore, 1881–1963) von Kármán's equation **(14.19)**

$$\lambda = K_0 \left| \frac{dv}{dr} \middle/ \frac{d^2v}{dr^2} \right|,$$

$$K_0 \simeq 0 \cdot 38. \tag{14.58}$$

K_0 had then been found to be the above constant for turbulent systems with (Osborne, 1842–1912) Reynolds numbers **(14.20)** between 2×10^3 and $1 \cdot 6 \times 10^6$.

Recall that the Reynolds number is defined as

$$\text{Re} = \frac{\varrho v L}{\eta}, \tag{14.59}$$

where L is a length of the dimensions of the system. Re is a measure of the ratio of the inertial to viscous forces in a system, and for $\text{Re} > \text{Re}_{\text{critical}} \simeq O(100)$ there will be turbulence **(14.21)**. Thus, since the solar nebula would have $\text{Re} \gg 10^6$ (see below), the system would of course be turbulent, although one probably would have to modify the value of K_0. Still, using eqn. (14.58) gives

$$\lambda = 0 \cdot 26r. \tag{14.60}$$

To compare this to the prediction of the von Weizsäcker theory note that λ corresponds to $(r_{n+1} - r_n)$ and that r corresponds to $\frac{1}{2}(r_{n+1} + r_n)$. Thus, using eqn. (14.52),

$$\lambda = 2e_{\text{max}}r, \tag{14.61}$$

since

$$(r_{n+1} - r_n) = e_{\text{max}}(r_{n+1} + r_n). \tag{14.62}$$

From Table 14.1 we see that the $2e_{\text{max}}$ in eqn. (14.61) is of the same order as the number in eqn. (14.60), which Tuominen found encouraging. Since von Kármán's equation had not been verified in systems where such large eddies could be formed, Tuominen suggested that there was a difference of order 2 in the numerical coefficients of eqns. (14.60) and (14.61) because K_0 probably had to be modified in very large systems. Tuominen also noted that both the von Kármán and von

Weizsäcker theories predicted eddy sizes in the solar nebula that increased linearly with distance.

Tuominen's arguments were taken up by ter Haar (14.14), who pointed out that if one considers just a geometric progression, then the heavier parent-body planetary and Jupiter systems have larger geometric progression ratios than those systems with the smaller parent-bodies (the Saturn and Uranus systems). This suggested to him that in cases where gravitation was less important the progression ratio would be closer to that given from eqn. (14.60), i.e. a ratio due to turbulence considerations alone. Contrariwise, we note that this idea also implies that when gravitation is more important the progression ratio should have a limiting value of some number like 1·73 [i.e. from eqns. (14.52) and (14.54)].

Continuing along these lines, ter Haar (14.22) later considered the implications of the more sophisticated theory of turbulence developed by Heisenberg (14.23–14.25). The first step was to assert that the solar nebula would roughly have[†]

$$\varrho \sim 10^{-10} \text{ g/cm}^3,$$
$$v \sim 10^5 \text{ cm/sec},$$
$$L \sim 10 \text{ AU} = 1\cdot5\times10^{14} \text{ cm}, \tag{14.63}$$
$$\eta \sim 10^{-2} \text{ poise}.$$

Then, by eqn. (14.59),

$$\text{Re} \sim 1\cdot5\times10^{11}, \tag{14.64}$$

so that there would be turbulence. Thus, the pressure in the nebula really would consist of two parts, the gas and turbulent pressures:

$$P = P_{\text{gas}} + P_{\text{turb}}. \tag{14.65}$$

The turbulent pressure is

$$P_{\text{turb}} = \frac{1}{3}\varrho u^2 \equiv \frac{1}{3}\varrho f v_K^2 = \frac{1}{3}\varrho f\left[\frac{GM}{\mathscr{R}}\right]. \tag{14.66}$$

[†] In eqn. (14.63), the value of v corresponds to a distance of $\approx$900 AU, and the value of η is one that is usually associated with a liquid. Better values would have been $v \approx 7\cdot6\times10^5$ cm/sec ($r \approx 20$ AU) and $\eta \approx 10^{-4}$ poise (roughly that of a hydrogen gas). But then the argument given by eqn. (14.64) would have been even better.

We will show later that the last part of the above equation defines f to be roughly given by

$$f \sim 1/3. \tag{14.67}$$

For the temperature in the nebula, ter Haar found

$$T = w(\mathcal{R})\Gamma\mathcal{R}^{-1/2},$$
$$\Gamma = (400°\text{K})\,(\text{AU})^{1/2}, \tag{14.68}$$
$$w(0{\cdot}4\text{ AU}) = 0{\cdot}9, \qquad w(5\text{ AU}) = 0{\cdot}1,$$

where $w(\mathcal{R})$ represents the effect of a finite optical length in the disk. The pressures can thus be compared as

$$\frac{P_{\text{gas}}}{P_{\text{turb}}} = \frac{w(\mathcal{R})\Gamma k\mathcal{R}^{1/2}}{fGMm_p\mu} \sim 0{\cdot}002\left(\frac{\mathcal{R}}{1\text{ AU}}\right)^{1/2}. \tag{14.69}$$

We see that in the solar disk P_{turb} is always at least a hundred times larger than P_{gas}, so P_{gas} can be ignored. Thus,

$$P = P_{\text{turb}} = \frac{1}{3}f\varrho\,\frac{GM}{\mathcal{R}}, \tag{14.70}$$

so that the equilibrium equations similar to eqns. (14.34) and (14.35) that come from eqn. (14.1) are

$$\frac{f}{3}\frac{d\ln\varrho}{dr} = r\left[-\frac{1}{\mathcal{R}^2}+\frac{f}{3}\frac{1}{\mathcal{R}^2}+\frac{\mathcal{R}\omega^2}{GM}\right], \tag{14.71}$$

$$\frac{f}{3}\frac{d\ln\varrho}{dz} = z\left[-\frac{1}{\mathcal{R}^2}+\frac{f}{3}\frac{1}{\mathcal{R}^2}\right]. \tag{14.72}$$

Equations (14.71) and (14.72) are easier to solve than eqns. (14.34) and (14.35) because now the gradient of the pressure has the same functional form as the gravitational force, with a size about one order of magnitude smaller. This pair of equations is solved just as was the previous pair. Integrating eqn. (14.72) we have

$$\ln\varrho = \left(1-\frac{3}{f}\right)\ln\mathcal{R}+\tau(r). \tag{14.73}$$

Putting this into eqn. (14.71) then gives

$$\frac{d\tau(r)}{dr} = \frac{3}{f}\frac{\mathcal{R}r\omega^2}{GM}. \tag{14.74}$$

Since $\tau(r)$ still has to be independent of z and since the pressure gradient is still small compared to the gravitational force, we can write

$$\omega^2 = q^2(r)\frac{GM}{r^3}\frac{r}{\mathcal{R}},$$

$$q^2(r) \equiv 1 - \gamma r. \tag{14.75}$$

$q^2(r)$ describes the first order deviation in the pressure or density [see eqn. (14.71)] gradient due to dissipation and it yields a Kepler orbit in the ecliptic when $\gamma \to 0$. Combining eqns. (14.73)–(14.75) gives

$$\varrho = \varrho_0\left(\frac{\mathcal{R}}{\mathcal{R}_0}\right)\left(\frac{r}{\mathcal{R}}\right)^{3/f}e^{-r(3\gamma/f)}. \tag{14.76}$$

In the equatorial plane this is

$$\varrho_e(r) \equiv \varrho(r, z = 0) = \varrho_m\left(\frac{r}{r_m}\right)e^{(1-r/r_m)}, \tag{14.77}$$

where ϱ_m is the maximum value of the density in the equatorial plane:

$$\varrho_m \equiv \varrho(r = r_m \equiv f/3\gamma, z = 0) = \frac{\varrho_0}{\mathcal{R}_0}\frac{fe^{-1}}{3\gamma}. \tag{14.78}$$

One can easily verify that $\varrho_m = 10^{-10}$ g/cm^3 and $r_m = 3\times10^{13}$ cm gives a total mass in the solar nebula of 0·1 M. The height (h) of the disk is given by

$$\frac{\varrho(r, h)}{\varrho(r, 0)} = \frac{1}{2},$$

$$\frac{h}{r} = [4^{f/(3-f)} - 1]^{1/2} \simeq 0\cdot36. \tag{14.79}$$

ter Haar studied other properties of this nebula, such as its lifetime and the properties it predicts for the planets and satellites. But at first the most important question to us is the scale size of the turbulence in this disk. This is what must give us a Titius–Bode Law, and also yield the form of eqns. (14.69) and (14.70). This was investigated in a separate appendix **(14.26)**.

As in Heisenberg's theory **(14.23–14.25)**, one starts by considering ε_k, the rate of dissipation of energy from all eddies with wave numbers less than a given k (from all eddies with sizes greater than a given $\lambda = 2\pi/k$) to all eddies with wave numbers greater than this k (to all eddies with sizes less than this λ). If $E(k)$ is the spectrum of turbulence and $v = \eta/\varrho$ is the "kinematic viscosity", Heisenberg proposed that in addition to the usual thermal contribution to ε_k there is a similar "mechanical" dissipation; i.e.

$$\varepsilon_k = \varepsilon_k(\text{thermal}) + \varepsilon_k(\text{mechanical}), \tag{14.80}$$

$$\varepsilon_k(\text{thermal}) = \varrho v \,\overline{|\text{curl } \mathbf{v}|^2} = 2\varrho v \int_0^\infty E(k')k'^2 \, dk', \tag{14.81}$$

$$\varepsilon_k(\text{mechanical}) \equiv 2\varrho v_k \int_0^\infty E(k')k'^2 \, dk'. \tag{14.82}$$

It is assumed that the new "eddy kinematical viscosity" (v_k) is given by

$$v_k = \varkappa \int_k^\infty \frac{dk'}{(k')^{3/2}} [E(k')]^{1/2}, \tag{14.83}$$

$\varkappa$ being a numerical constant of order one.

Given that there is some largest eddy, the double integral in eqns. (14.82)–(14.83) involves two constants of integration. One constant will be related to the largest eddy of the system (some k_0). The other constant, $E(k_0)$, is related to the amount of energy given to this largest eddy since one assumes that the driving force transfers most of the energy to the largest eddies which in turn dissipate it to the smaller eddies. Note that with this description,

$$E(k) = 0, \qquad k < k_0. \tag{14.84}$$

Since the driving force will come from differential rotation, a model can be made for the left-hand side of eqn. (14.80), i.e. the amount of energy that is given to the system. This is

$$\varepsilon_k = \frac{1}{2}\varrho\xi\frac{d}{dr}\left| l(k, k_0)\, u(k, k_0)\, \frac{dv_K^2}{dr} \right|. \tag{14.85}$$

Here v_K represents the mean mass velocity and ξ is a number of order unity. l and u are again the mixing length and eddy velocity, but now they have been analyzed into components. This is to give the energy transferred by the driving force into eddies with wave numbers between k_0 and k (sizes between λ_0 and λ). This energy is then dissipated into the smaller eddies [the right-hand side of eqn. (14.81)].

By making the same type of assumption as Heisenberg did for ε_k(mechanical) in eqn. (14.83), one has for eqn. (14.85)

$$\varepsilon_k = \varrho\beta\varkappa \int_{k_0}^{k} \frac{dk'}{(k')^{3/2}}\, [E(k')]^{1/2},$$

$$\beta \equiv \frac{1}{2}\xi\left| \frac{d^2v_K^2}{dr^2} \right|. \tag{14.86}$$

Combining eqns. (14.80)–(14.83) and (14.86) and differentiating by k,

$$\frac{v}{\varkappa} + \int_{k}^{\infty} \frac{dk'}{(k')^{3/2}}\, [E(k')]^{1/2} = \frac{1}{k^{7/2}[E(k)]^{1/2}}\left[\beta + \int_{k_0}^{k} E(k')k'^2\, dk' \right]. \tag{14.87}$$

This equation can be solved. In particular, for fully developed turbulence ($\eta = v = 0$) the solution is

$$E(k) = \begin{cases} \dfrac{4}{3c^{4/3}}\dfrac{1}{k^{5/3}}, & k \geq k_0 = c\beta^{3/4}, \\[2mm] 0, & k < k_0, \end{cases} \tag{14.88}$$

c being a constant. But this is just the Kolmogoroff energy spectrum[†]

[†] The Kolmogoroff energy spectrum, which was independently obtained by Onsager (14.28) and von Weizsäcker (14.29), is the spectrum for eddies with large Reynolds numbers. The Heisenberg theory is an extension that includes eddies with small Reynolds numbers, and so has the Kolmogoroff spectrum as a limit.

(14.27). When written in the standard way it is

$$E(k) = E(k_0) \left(\frac{k_0}{k}\right)^{5/3}, \qquad k \geq k_0, \qquad (14.89)$$

$$\overline{u^2} = 2 \int_0^\infty E(k) \, dk = 3E(k_0) \, k_0. \qquad (14.90)$$

Combining eqns. (14.88)–(14.90) and then using eqns. (14.86) and (14.56), we get

$$\overline{u^2} = 4\beta/k_0^2 = 4\xi v_K^2/r^2 k_0^2. \qquad (14.91)$$

But since $\sqrt{\overline{u^2}}$ represents the average eddy velocity and v_K represents the driving Kepler velocity, we expect that

$$\overline{u^2} \sim f v_K^2,$$
$$f = O(\tfrac{1}{3}). \qquad (14.92)$$

Then eqns. (14.91) and (14.92) are consistent if

$$\lambda_0 \equiv \frac{2\pi}{k_0} = \pi \left(\frac{f}{\xi}\right)^{1/2} r \equiv r\delta, \qquad (14.93)$$

δ being of order unity.

Since λ_0 represents the largest-sized eddy, where most of the energy will be deposited, the idea is that the structure of the turbulent nebula at any radius will be determined by eqn. (14.93). Actually, of course, there can be no discontinuous spectrum such as eqn. (14.88), as is discussed in Chapter 6 of Batchelor **(14.16)**. However, in this theory of the Titius–Bode Law, one proposes that the approximations we have made (such as $\eta = 0$) are valid enough in a statistical sense to give the dominance predicted by eqn. (14.93). Then eqn. (14.93) is the justification for the von Weizsäcker theory of the Titius–Bode Law and δ in eqn. (14.93) represents the $2e_{\text{max}}$ of eqn. (14.61).

The appeal of this idea is that to obtain the Titius–Bode Law by the theory of von Weizsäcker one is using the physical argument that in statistically turbulent media the energy transfer will be dominated by the largest vortex cells in a particular region. That is, eqn. (14.93) is valid and the eddy sizes increase linearly with distance. Given this, one argues that rings of vortex cells will be set up, roughly as envisioned by von Weizsäcker, because such regular rings will minimize energy dissipation. Now of course there will be a distribution of eddy sizes, but at a particular radius (or ring) the size given by eqns. (14.61) and (14.93) will dominate to yield the Law.

However, even giving that all of the other difficulties in this picture can be taken care of and that a regular turbulence pattern can be set up, there still remains the problem of maintaining this large-scale turbulence for a long enough period of time in the first place.

If the energy for the turbulence indeed comes from differential rotation, then the time available for condensation is no longer than about 1000 years (14.22) before the nebula is dissipated (part of it falling onto the Sun, part of it going to infinity). This is very short, and if this picture also held for the interstellar gas of the galaxy, then it would have dissipated in a time short compared to the age of the galaxy (8.7), something which clearly has not happened.

The above bound for the dissipation time can be understood. First realize that the total energy dissipated per unit volume per unit time is

$$\frac{d^2E}{dV\,dt} = \varepsilon_{k \gg k_0} \rightarrow \text{const. (independent of } k\text{).} \qquad (14.94)$$

For the case of $v = \eta = 0$, eqn. (14.94) can be evaluated explicitly by combining eqns. (14.80)–(14.83) and (14.89)–(14.90) to give

$$\frac{d^2E}{dV\,dt} = \frac{\sqrt{3} \cdot \varkappa}{8} \varrho(\overline{u^2})^{3/2} k_0. \qquad (14.95)$$

From eqns. (14.91)–(14.93) this can be put in the form

$$\frac{d^2E}{dV\,dt} = \frac{2\pi \sqrt{3} \cdot \varkappa f^{3/2}}{8\delta} \frac{\varrho v_k^3}{r}. \qquad (14.96)$$

To evaluate this, we make the approximation that between the distances of $0{\cdot}36r$ above and below the equatorial plane the density is given by its value in the equatorial plane [see eqn. (14.79)], and it is zero elsewhere.

120

Then

$$\varrho(r, z) = \varrho_e(r)\,\theta(0{\cdot}36r - |z|), \qquad (14.97)$$

where θ is the unit step function. Combining eqns. (14.96), (14.97), (14.77), and (14.56) and integrating over dV gives

$$\frac{dE}{dt} = \frac{(0{\cdot}72)\pi^{5/2}\sqrt{3}\cdot e\varkappa f^{3/2}}{8\delta}\,[\varrho_m(GM)^{3/2}r_m^{1/2}]. \qquad (14.98)$$

Using the virial theorem, the total kinetic energy can be similarly found to be

$$E_0 = \frac{1}{2}\int \frac{\varrho MG}{r}\,dV = (0{\cdot}72)\,e[2\pi r_m^2 GM], \qquad (14.99)$$

so that the decay time is given by

$$\tau = \frac{E_0}{dE/dt} = \frac{8\delta}{\sqrt{3}\cdot\varkappa\pi^{5/2}f^{3/2}}\,[2\pi r_m(r_m/GM)^{1/2}]. \qquad (14.100)$$

From the discussion after eqn. (14.78), the quantity in the brackets is $(2)^{3/2}$ years (the Kepler period of revolution at the distance of $r_m = 2$ AU). If one takes $\varkappa = 0{\cdot}85$, $f = 1/3$, and $\delta = 0{\cdot}518$ [see eqn. (47) of **(14.22)**, eqn. (14.92), and eqns. (14.93) and (14.61), respectively], then

$$\tau = 2{\cdot}4 \text{ years.} \qquad (14.101)$$

If one took the most optimistic upper limits for $(dE/dt)^{-1}$ and E_0 [ter Haar **(14.22)** used values ~ 30 and 6 times higher than ours], this figure might be increased to ~ 1000 years. This number would allow the creation of rocks and ices that might remain after the gas left, but it certainly would not be enough for the building up of planets which would take $\sim 10^7$ years.

Thus, the acceptability of a von Weizsäcker type mechanism for the Titius–Bode Law depends critically on a way being found for the disk to last much longer than the kind of time given by eqn. (14.100). We shall return to this point in Chapter 15.

D. Kuiper's Theory

Just before ter Haar and Chandrasekhar proposed their later modification of von Weizsäcker's theory, Kuiper published the first (14.30) of a series of articles (14.30–14.32) which resulted in a counter proposal to von Weizsäcker's theory.

Kuiper suggested that the masses and distances of the planets are related in an intimate way. The mechanism for this was the condensation of large, proto-planet, gas clouds from the nebula that were gravitationally bound despite the disruptive effects of tidal forces from the Sun. In his theory Kuiper took the picture of a turbulent nebular disk that was near the Roche Limit (see Chapter 10). In fact, he proposed that the condensation into gas clouds occurred at the Roche Limit and that it was the pressure effects of turbulence that allowed particular vortices, once formed, to pass the Roche Limit and to become separate, gravitationally bound, gas, proto-planets. In effect, then, Kuiper pictured spheres of stability, not rings of stability, from which the planets came.

To obtain an estimate of what sort of relationship for the planets' masses and distances such a mechanism would imply, consider eqn. (10.4) for the Roche Limit. As we mentioned in Chapter 10, the number on the right-hand side varies depending on the type of system that is discussed, but it is of order unity. [See the discussion in ref. (6.21) and in appendix A of ref. (14.32).] If we cube eqn. (10.4), we have for the solar system

$$\frac{m}{M}\frac{r^3}{(\frac{1}{2}\varDelta)^3} = \frac{\varrho(\text{planet})\,(\frac{1}{2}\varDelta)^3}{\varrho(\text{Sun})\,a^3}\frac{r^3}{(\frac{1}{2}\varDelta)^3} \geqslant O(1), \qquad (14.102)$$

where we identify $\frac{1}{2}\varDelta$ as being the radius of the gas cloud when it broke from the nebula.

If we accept that there would be gas spheres competing with this cloud both from the inside and from the outside and state that these two spheres also condensed just at the Roche Limit, then there will be a relationship like eqn. (14.102) for adjacent pairs. Specifically, we then have

$$\left(\frac{m}{M}\right)\left(\frac{r}{\Delta}\right)^3 \sim O(1),$$

$$r \equiv \tfrac{1}{2}(r_{n+1}+r_n),$$ (14.103)

$$\Delta \equiv (r_{n+1}-r_n),$$

where m is the average mass for adjacent pairs of planets. Thus, for two adjacent planets or satellites we have a mass–distance relation.

Kuiper then claimed, on intuitive grounds, that there are two factors that modify eqn. (14.103):

1. Since there originally was a disk nebula instead of a spherical nebula, the proto-planets' masses would be smaller by a factor of about 10^{-1} to 10^{-2}.

2. Since only about 1% of the mass of the nebula condensed into the planets, another factor of 10^{-2} has to be inserted.

Thus, Kuiper's final relation for the solar system is

$$\left(\frac{m}{M}\right)\left(\frac{r}{\Delta}\right)^3 \sim 10^{-4}.$$ (14.104)

A plot of the experimentally observed masses–distances is given in Fig. 14.2. Note that even given the broad assumptions, the curve of eqn. (14.104) fits the data very roughly and the top plots (for groups of pairs of planets with approximately the same ratios of masses) follow a different curve. Empirically this latter curve is

$$\log\left(\frac{m}{M}\right) = 2\cdot5 \log\left(\frac{\Delta}{r}\right)^3 - A.$$ (14.105)

Equation (14.105) was found to be good with A being 1·8 for planets of comparable masses and 3·6 for planets of very unequal masses, the same Law holding for satellites. Kuiper tried to explain the differences between eqns. (14.104) and (14.105) by arguing that different ratios of adjacent masses meant that the original proto-clouds had retained different fractions of their original masses and so there was a change in slope.

This being done, Kuiper then took the rough agreement of eqns.

123

(14.104) and (14.105) as proof that the planets indeed had condensed at the Roche Limit by such a mechanism. However, with the assumptions that Kuiper made, his theory was, at this point, at best a parameterization with a rough physical interpretation.

After the work mentioned in the previous section became known to him, Kuiper expanded his ideas (**14.31, 14.32**). Recall that the ter Haar and Chandrasekhar result (**14.22, 14.26**) predicted a Kolmogoroff

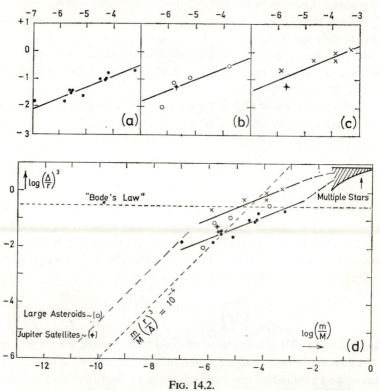

FIG. 14.2.

Kuiper's (m/M) vs. $(\Delta/r)^3$ relationship for the planets and the satellites. Abscissas are $\log(m/M)$ and the ordinates are $\log(\Delta/r^3)$. (a) Pairs of planets or satellites of nearly equal masses (ratio less than 5); (b) same, but mass ratio between 5 and 10; (c) same, but mass ratio more than 10; (d) combined plot. The data are explained in ref. (**14.30**).

spectrum for the turbulence eddies. Although this system gives a largest eddy size in agreement with the von Weizsäcker picture [eqn. (14.93)], Kuiper did not accept the assertion that the primary large eddies would dominate the form of the system, even though most of the energy would be transferred to them.

Contrariwise, Kuiper pointed out that from the von Weizsäcker (14.29) discussion of the Kolmogoroff spectrum,

$$u(k) \propto k^{1/3}. \tag{14.106}$$

Thus, the energy spectrum $[E(k) \, dk]$ can be written in terms of the eddy-size distribution function $[n(k) \, dk$ or $n(\lambda) \, d\lambda]$ as

$$E(k_0) \left(\frac{k_0}{k} \right)^{5/3} dk = E(k) \, dk \propto u^2(k) \, n(k) \, dk. \tag{14.107}$$

Combining eqns. (14.106) and (14.107) we have

$$[n(k) \, dk] = \text{const. } k^{-1} \, dk, \tag{14.108a}$$

or, after changing variables,

$$[n(\lambda) \, d\lambda] = \text{const. } \lambda^{-1} \, d\lambda. \tag{14.108b}$$

Kuiper asserted that this distribution would probably not align itself in a regular pattern, even for short periods, and he schematically suggested the distributions shown in Fig. 14.3a, b as two of many possible alternatives. (These distributions were admittedly schematic and should not be taken as valid, descriptive representations. We show them anyway because they have become part of the "folklore" of the Titius–Bode Law.) Thus, as he did not see why a regular system must be formed and because he objected to the short lifetimes of the secondary "roller-bearing eddies" (these were to be the condensation centers), Kuiper felt the von Weizsäcker picture had to be abandoned. (Remember, however, that in the von Weizsäcker picture the formation of large pieces of matter can take place after the vortices have died out, by accretion of the rocks and ices that were formed at the interfaces of the rings.)

As an alternative, Kuiper proposed an elaboration of the picture that we discussed at the beginning of this section. The elaboration was on the description of the gravitationally bound clouds.

Remember that in his first theory it was assumed that the proto-planets condensed when the solar nebula was near the Roche Limit. The increased pressure in a turbulence cell would randomly allow it to exceed the Roche Limit and remain gravitationally bound. The only question, then, was if such a cloud would remain bound.

Kuiper answered this by using the properties of *gravitational instability* as described by Jeans **(14.33)**. The instability depends upon whether a gas cloud by contracting will have less energy (sum of addi-

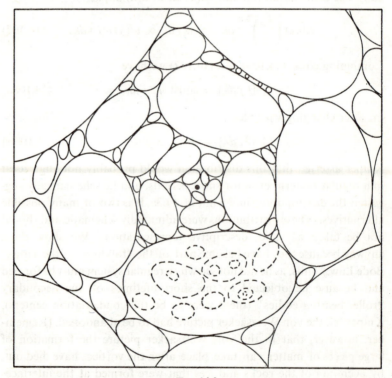

FIG. 14.3a.

Kuiper's schematic distribution of eddies simulating the Kolmogoroff spectrum. The Sun is at the center, and "interior" eddies are shown in one instance.

tional kinetic energy minus the potential energy) and be unstable to further contraction, or if it will gain energy and thus be stable. The criterion is instability (stability) if its dimension λ has $\lambda > \lambda_S$ ($\lambda < \lambda_S$) where

$$
\lambda_S = \left[\frac{\pi c_P v_{KE}^2}{3 c_V G \varrho} \right]^{1/2}
$$
$$
= 0 \cdot 05 \left(\frac{\varrho(\text{Roche})}{\varrho} \right)^{1/2} r, \tag{14.109}
$$

and c_P and c_V are the specific heats at constant pressure and volume of the gas. (The last equality comes from evaluating Kuiper's model.)

Thus, Kuiper said that turbulence can be thought of as providing the initial density fluctuations, and gravitational instability as amplifying them (if $\lambda > \lambda_S$) or damping them (if $\lambda < \lambda_S$). Therefore, the

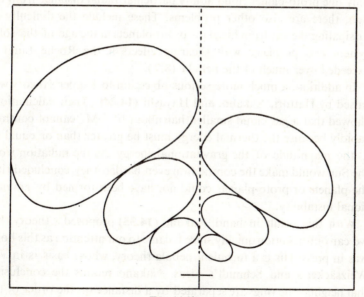

FIG. 14.3b.

Kuiper's representation of two alternative arrangements of eddies if the largest-sized eddies, at any one distance, prevail.

127

amplification will be greatest for the maximum-sized eddies given by $\lambda_0 > \lambda_S$, and this was why there is a rough agreement with eqn. (14.93), i.e. the Titius–Bode Law for λ_0.

In this final picture the amplification, and hence the number of condensations of various sizes, depends critically on the density of the nebula. So, Kuiper concluded that the Titius–Bode type Law of eqn. (14.104) is really a function of the original solar nebular density. This entails taking into account the thickness of the disk, which Kuiper pointed out is much thinner than is often envisioned because one has to consider the effect of the nebula's gravitational field on itself in the z direction.

Kuiper's observations did offer valid criticisms of von Weizsäcker type theories. But they did not themselves yield a theory that was amenable to a rigorous and critical interpretation. In addition to the objections we have previously mentioned and the question of why there was only one proto-planet cloud at any particular radial distance from the Sun, there are also other problems. These include the difficulty of dissipating the gas from his giant proto-planets in the age of the solar system and problems with shearing effects if the Roche Limit is exceeded over much of the nebula (8.7).

In addition, a much more serious objection to Kuiper's theory was raised by Hattori, Nakano, and Hayashi (14.34). Their calculations showed that a gas cloud smaller than about $10^{-2} M_\odot$ cannot contract rapidly because the thermal energy must be greater than or equal to $\frac{1}{2}$ the magnitude of the gravitational energy. As the radiation from the Sun would make the contraction even harder, they concluded that the planets or proto-planets could not have been formed by gravitational instability.

With this result in hand, Nakano (14.35) proposed a theory that we can only mention quickly since it came to our attention as this book was in press. His is a turbulent nebula theory, whose basis is in von Weizsäcker's and Schmidt's ideas. Nakano retains the conclusion that neighboring rings are separated by a distance on the order of the largest-sized eddies (i.e., scale size), so he obtains the same Titius–Bode Law as von Weizsäcker. However, what is different is the means for the separation into rings, it coming from two highly speculative assumptions: (1) The mean turbulent velocity at r is taken as being proportional to the Kepler angular velocity, so that the average angular momentum per unit mass is

$$\langle u \rangle = (GMr)^{1/2}. \qquad (14.110)$$

The variation of u with r will cause a drift of matter, leading to a density fluctuation. (2) Then it is assumed that this density fluctuation, along with the quicker decay of turbulence near the (in hydrostatic equilibrium) Sun, will allow successive rings to be formed with time, each being further and further out from the Sun.

So, we must close this chapter with the observation that although much insight has been obtained from nebular theories into the origin of the solar system and into the Titius–Bode Law, no theory has been given for the geometric progression which as of yet is able to answer all criticisms. However, in light of the recent advances that have been made in this area, it is to be hoped that the viable parts of the theories we have discussed contain the clues which will allow this problem to be solved.

CHAPTER 15

Conclusion

A. Review

In the preceding chapters we have reviewed the interesting history and theory of the Titius–Bode Law. We saw how, with an intellectual heritage going back to the work of Kepler, during the seventeenth and eighteenth centuries ideas developed that there had to be some sort of relationship among the distances of the various planets and that, indeed, there might even be missing planets.

Upon this foundation, Johann Daniel Titius von Wittenberg used his 1766 translation into German of Charles Bonnet's *Contemplation de la Nature* as a vehicle to first promulgate what became known as "The Titius–Bode Law". After the second edition of this translation was seen by Johann Elert Bode in 1772, Bode took the Law as his own and incorporated it into the second edition of his *Anleitung zur Kenntniss des gestirnten Himmels*.

As we discussed in Chapter 3, Bode eventually received the credit for the Law due to (a) Bode's fame, (b) his continued use of the Law, and (c) the respective places where he versus. Titius published it. The Law's checkered history reached a high point when it predicted the locations of Uranus and the asteroid belts and was somewhat success-fully applied to the satellite systems of the major planets. However, after the disagreement of its predictions with the distances of Neptune and later Pluto, the Law became a "black sheep" in astronomy. Astronomers thereafter were in disagreement as to the Law's possible significance.

The 150 years following the Law's discovery saw many reformula-tions in attempts to increase its numerical agreement with the planetary distances and to widen its applicability to the satellite systems of the major planets. These formulations culminated in those of Blagg and Richardson, which showed that the best way to represent the Law was without the *ad hoc* first term and by using a geometric progression of 1·73 (*not* the 2·0 of the original formulation) multiplied by a periodic

130

function. With this description, the agreement of the Law with observation was impressive. Among other things, the disastrous troubles with the distances of Neptune and Pluto were removed.

Although rough attempts were made to give a physical interpretation to the Law in the last century, it was only in this century that serious theories were actually proposed. Reviewing these efforts in the perspective of our present knowledge about the solar system, we first came to the conclusion that during its history the solar system had three periods: (I) *the disk period*, (II) *the period of aggregation*, (III) *the planet period*. Then we proposed that (a) the geometric progression is due to a fluid and/or magnetohydrodynamical mechanism which occurred during the disk period, and (b) the "evolution" or periodic function of the Law comes from a tidal or point gravitational relaxation that took place during the planet period. We pointed out, however, that our reasoning is not ironclad. Therefore, we kept before the reader the conflicting viewpoint that the entire Law might have been caused by gravitational and/or tidal encounter mechanisms that took place during the planet period.

B. Outlook

Given this state of affairs, there certainly remains much interesting work to be done on this problem, both on the historical as well as on the scientific sides.

Even though our discussion in Chapters 2–4 has cleared up the main questions on the roles of Bonnet, Titius, and Bode in the discovery of the Law, many of the exact details and perhaps further surprising revelations remain hidden. It would be a very recommendable project for the interested scholar to search for available papers of the three men. With diligence he might find the letters that must have been exchanged among them, and we could then gain a surer insight into the history we described in Chapters 2–4.[†]

[†] Prime places to begin such a search for the papers of these men would be at the University Library in Neuchâtel (Bonnet), the Wittenberg–Halle University and "Land" Library in Halle (Titius), and the Archives of the Deutsche Akademie der Wissenschaften in Berlin (Bode).

On the scientific side, we hopefully will soon be able to obtain at least partial solutions to the questions we have posed in the text.

The first point would be to correctly generalize Hills' idea and numerically calculate the evolution of a three-dimensional model of the solar system using the real masses of the planets. If, in a time corresponding to the age of the solar system, such a model could evolve from distances given by a progression of 1·73 to positions closely approximating their present distances, but could not do this from arbitrary initial distances, then one would have evidence for a gravitational relaxation being the cause of the "evolution function" of the Titius–Bode Law. Contrariwise, if the model could evolve from arbitrary distances it would be evidence for a gravitational origin for the entire Law, or if the model could not evolve very much at all one would then have to doubt such a mechanism for any part of the Law.

However, even if the above calculations were done, this would still not be a complete discussion since, as mentioned in Chapter 13, such a calculation is beset by the uncertainties that tidal effects would produce. These effects also should be calculated since there are reasons to believe that tidal forces could be more important in the "evolution". This would especially hold during the early part of the planet period. Then the planetary atmospheres would be larger and the planets would be less compressed so that tidal dissipation of energy would be greater. Thus, one should perform calculations similar to those of the preceding paragraph to see what kind of evolution tidal forces could produce —and best of all, calculations to see what effects would be caused by a combination of gravitational and tidal forces.

To summarize: The (preferably combined) gravitational and tidal evolutions of a model, three-dimensional, solar system with realistic masses and time scales should be calculated. It should be discovered if such evolutions are able to describe the relaxation to the present planetary distances (a) at all, (b) from arbitrary initial distances, or (c) from distances corresponding to a geometric progression of 1·73.

This last possibility would imply the separate origin of the geometric progression of the Law for which we have argued. Then, given that we must look elsewhere for the origin of the geometric progression, we are constrained to discuss a nebular mechanism, as mentioned. In particular, for a von Weizsäcker type nebular theory, we saw in Chapter 14 that the big problems have to do with (a) the lifetimes of the turbulence eddies in the nebula and (b) with the necessity that the largest-sized eddies of the Kolmogoroff turbulence spectrum dominate over the rest of the vortex cells to form regular circulation and condensation patterns. It is our opinion that partial insights can be given into both of these problems. For example, the first of them should be re-examined in the context of Hoyle's ideas.

Recall that in the von Weizsäcker theory the nebular mass was about 10% of the solar mass with approximately 1% of the nebula eventually forming the planets. The nebula was viewed as having existed at its largest extent from the time of formation. Hoyle, on the other hand, had the nebula being 1% of the solar mass of which 10% became the planets so that, of course, the same amount of matter became the planets in both theories.

But more important, contrary to von Weizsäcker, Hoyle had the nebula growing outward from an initial position of approximately 0·2 AU, where it had formed due to the rotational instability of the contracting Sun. Energy and angular momentum were transferred to it from the Sun by the magnetic brake mechanism. If we put this aspect of Hoyle's theory into our picture of the nebula, we can see from our discussions in Chapters 9 and 14 that enough energy per unit time would be transferred to the nebula to maintain the turbulence for a period of approximately 10^7 years. Most of the expansion would happen during the beginning of this period so that turbulence would be maintained for a long time without much relative increase in the size of the nebula.

However, we must point out that this concept is harder to defend when applied to the satellite systems. There the magnetic brake mechanism—if operative—should not and (since the satellites are close to the mother planets) obviously did not have such a strong effect on the evolution of the planetary nebulas. Some nebular turbulence and energy transfer must have existed, but whether it was enough to have allowed large condensations in these systems cannot be convincingly proved.

As for the second problem in a turbulence theory, that of creating a regular circulation system dominated by the largest-sized eddies in the turbulence, we point out that it is not inconceivable that such an event could occur. In fact, there is a well-known example close at hand—the Earth's lower atmosphere **(15.1)**.[†]

In both hemispheres of the Earth's atmosphere there are two toroidal, mass circulation systems. Taking the northern hemisphere for definiteness, there is a circulation system that goes from near the equator to about 30°N.–40°N. composed of ascending warm air in the south and descending cooler air in the north. This is the (George, 1685–1768) Hadley Cell of the tropics. Further north there is a weaker meridional circulation in the opposite sense. It ascends at about 65°N.–75°N. and descends near 45°N. and is called the (William, 1817–91) Ferrel Cell.

When these cells were studied in detail, it became clear that if actually just these axially symmetric circulations existed on the rotating Earth, then there would have been very large zonal wind speeds in the connecting meridional branches of the flow. The existence of the cyclonic and anticyclonic eddies in weather systems was viewed as the way these excessive wind patterns were avoided.

At the same time it was pointed out that these eddies could accomplish a great deal of the meridional heat and mass exchange. Later investigations showed that this is true. In fact, except for the regions nearest to the equator, it was found that eddy processes predominate in the meridional exchange of momentum. Finally, it was concluded that although this exchange is done by small-scale turbulence in the lowest levels, in the higher latitude middle troposphere the momentum transfer is probably dominated by synoptic-scale eddies (i.e. eddies the size of the general circulation system).

Now we by no means wish to imply that the Earth's lower atmo-

[†] For an excellent and thorough description of circulation in the Earth's lower atmosphere (pressures greater than 100 mb), the reader is referred to ref. **(15.1)**, from where our discussion is taken.

sphere is the same system as the solar nebula. What we do intend to emphasize is that this shows a regular circulation system can be set up by turbulence whose largest-sized eddies are the size of the circulation pattern; i.e. the atmosphere is used only as an example to demonstrate that von Weizsäcker type turbulent mechanisms for the geometric progression of the Titius–Bode Law cannot be disregarded out of hand for being contrary to common sense. Such mechanisms should be kept in mind as viable paths to investigate.

On the other hand, if one had to take the view that a turbulence mechanism could not be maintained, then one might be tempted to revert to the later ideas of Berlage. Recall that Berlage ultimately became convinced that although turbulence had existed in the early stages in the nebula, a regular Law could not have come from this and the turbulence must have quieted down. His concept was that in this quiet nebula there were forced perturbations due to hydrodynamical or magnetohydrodynamical effects, such as the solar wind. Then, if the nebula satisfied his "minimum energy loss" condition for the frictional dissipation due to the forced perturbations, it hopefully would have had an oscillatory mass distribution with the maxima (and hence preferred condensations) at distances that corresponded to a Titius–Bode type geometric progression.

Indeed, after eqn. (14.29) we showed that under the realistic conditions of radiative equilibrium and constant chemical composition in the nebula, applying Berlage's "minimum energy loss" concept yielded just such a progression. This kind of system is like a "forced vibrating plate", and if one had to accept that turbulence would die out and could not cause the geometric progression, then models of this type would constitute an alternative picture.

To summarize again: The main problems of a von Weizsäcker type geometric progression theory could be attacked by (a) using a magnetic brake energy transfer to maintain the turbulence for a long enough period of time to allow preferred condensation (although there are troubles applying this to the satellite systems), and by (b) recognizing that it is sometimes possible to have scale-sized eddies being the dominate transfer mechanism in a regular circulation system. However, if a turbulence mechanism must be abandoned, then something like Berlage's "minimum energy solution" for a forced, quiet nebula might give rise to "vibrating plate" periodic maxima in the nebular density. This would allow a preferred condensation at these particular radii and yield a Titius–Bode type geometric progression.

This being said, I lay my typewriter to rest, hoping that the entire subject of the Titius–Bode Law is now in a much more understandable condition than it was previously. However, as the reader can easily see, much more interesting work lies ahead before a *finis* can be written on this venerable (yet often damned) Law of planetary astronomy.

References

1.1 POGGENDORFF, JOHANN CHRISTIAN (1796–1877), *Biographisch-Literarischer Handwörterbuch zur Geschichte der exacten Wissenschaften* (Johann Ambrosius Barth, Leipzig, 1863), was consulted, along with some of the many later editions of this reference source. Reprinted by B. M. Israël, N.V., Amsterdam, 1965.

1.2 WILLIAMS, T. I. (ed.), *Biographical Dictionary of Scientists* (Adam & Charles Black, London, 1969).

1.3 IRELAND, N. O., *Index to Scientists* (F. W. Faxon Co. Inc., Boston, 1962).

1.4 DEBUS, A. G. (ed.), *World Who's Who in Science* (Western Publishing Co., Hannibal, Missouri, 1968).

1.5 TITIUS VON WITTENBERG, JOHANN DANIEL, translation into German from French: *Betrachtung über die Natur, vom Herrn Karl Bonnet* (Johann Friedrich Junius, Leipzig, 1766), p. 7.

1.6 BONNET, CHARLES, *Contemplation de la Nature* (Marc-Michel Rey, Amsterdam, 1764), p. 8.

1.7 GARDNER, M., "Some mathematical curiosities embedded in the solar system", *Scientific American*, **222**, No. 4, 108–12 (April 1970).

1.8 O'NEIL, W. M., *Fact and Theory, An Aspect of the Philosophy of Science* (Sydney University Press, Sydney, Australia, 1969), pp. 58–61.

1.9 GOOD, I. J., "A subjective evaluation of Bode's Law and an 'Objective' test for approximate numerical rationality", *J. Amer. Stat. Assoc.*, **64**, 23–49 (1969).

1.10 ANSCOMBE, F. J. (50–51); BROSS, I. D. J. (51–57); HARTLEY, H. O. (57–58); BARGMANN, R. E. (58–59); DAVID, H. A. (59); ZELEN, M. (59–60); ANDERSON, R. L. (60); DAVIS, M. (60–61), "Discussion of paper by I. J. Good, April 9, 1968", *J. Amer. Stat. Assoc.*, **64**, 60–61 (1969). The above are eight separate discussions of ref. (**1.9**).

1.11 GOOD, I. J., "Reply to the discussion", *J. Amer. Stat. Assoc.*, **64**, 61–66 (1969). This is a defense of ref. (**1.9**) against the criticisms of ref. (**1.10**).

1.12 HOUZEAU DE LEHAIE, JEAN CHARLES (1820–88) and LANCASTER, ALBERT-BENOIT (1849–1908), *Bibliographie Générale de l'Astronomie (ou Catalogue Méthodique des Ouvrages, des Mémoires et des Observations Astronomiques publiés depuis l'Origine de l'Imprimerie) jusqu'en 1880*, 2 vols. in 3 pts. Vol. 1 (F. Hayez, Brussels, 1887, 1889). Vol. 2 (Xavier Havermans, Brussels, 1882). Reprinted with additions and editing by D. W. Dewhirst (Holland Press, London, 1964).

1.13 *Royal Society Catalogue of Scientific Papers 1800–1900*, 4 Series in 19 vols. (Royal Society, London, 1867–1925).

1.14 KEMP, D. A., *Astronomy and Astrophysics, A Bibliographical Guide* (Macdonald & Co., Ltd., London, 1970).

REFERENCES

1.15 *Guide to Reprints 1971* (National Cash Register Company, Microcard Editions, Washington, D.C., 1970). Also see the same publisher's *Announced Reprints*, in Vol. **3**, as of (1971).

2.1 KEPLER, JOHANN, *Mysterium Cosmographicum* (Georgius Gruppenbachius, Tübingen, 1595).

2.2 CASPAR, MAX (1880–1956), *Bibliographia Kepleriana*, 2nd ed. (C. H. Beck'sche Verlagsbuchhandlung, München, 1968).

2.3 HOYLE, F., *Astronomy* (Macdonald & Co. Ltd., London, 1962), chap. 4.

2.4 KEPLER, JOHANN, *Astronomia Nova* (1609). No publisher was printed, but it undoubtedly was G. Vögelin, Heidelberg [see ref. (**2.2**), p. 47].

2.5 KEPLER, JOHANN, *Harmonices Mundi Libri V* [Ioannes Placvs, Lincii, Austriae (Johann Blancken, Lintz, Austria), 1619].

2.6 RHETICUS (Georg Joachim von Lauchen), *Narratio Prima (Ad Clarissimum Virum . . . de Libris Revolutionum . . . Copernici . . . Narratio Prima)* (Gedani, 1540); *ibid.*, 2nd ed. (Basileae, 1541). Translated into English by E. Rosen in *Three Copernican Treatises* (Columbia University Press, Morningside Heights, New York, 1939). See p. 147.

2.7 KOYRÉ, ALEXANDRE, *La Révolution Astronomique* (Hermann, Paris, 1961), pp. 138, 139 and 383. Also see ref. (**2.6**), Vol. I, pp. 55 and 105.

2.8 KEPLER, JOHANN, *Mysterium Cosmographicum*, 2nd ed. [Godefridi Tampachii, Francofurti (Frankfurt), 1621]. See also the translation into German by Max Caspar, *ibid.* (Dr. Benno Filser Verlag, Augsburg, 1923).

2.9 HOLTON, G., "Johannes Kepler's Universe: its physics and metaphysics", *Amer. J. Phys.* **24**, 340–51 (1956).

2.10 VON WOLF, CHRISTIAN FREIHERR, *Cosmologia Generalis* [Officina Libraria Rengeriana, Francofurti & Lipsiae (Frankfurt and Leipzig), 1731]. See pp. 46–47, 60–61.

2.11 VON WOLF(FEN), CHRISTIAN FREIHERR, *Vernünfftige Gedanken von den Absichten der natürlichen Dinge* (Rengerischen Buchhandlung, Frankfurt and Leipzig, 1723), §85. *Ibid.*, 2nd ed. (1726), misprinted as §84 on p. 139.

2.12 VON WOLF, CHRISTIAN FREIHERR, *ibid.*, 4th ed. (Halle, 1741), §85, p. 139.

2.13 KANT, IMMANUEL, *Allgemeine Naturgeschichte und Theorie des Himmels* (Johann Friederich Petersen, Königsberg and Leipzig, 1755), pp. 18 ff. and 163 ff.

2.14 LAMBERT, JOHANN HEINRICH, *Cosmologische Briefe über die Einrichtung des Weltbau* (Eberhard Kletts Wittib., Augsberg, 1761), p. 7.

3.1 BONNET, CHARLES, *Contemplation de la Nature*, contained in *Œuvres d'Histoire Naturelle et de Philosophie de Charles Bonnet*, Vol. IV (Samuel Fauche, Neuchâtel, 1781).

3.2 TITIUS VON WITTENBERG, JOHANN DANIEL, translation into German from French: *Betrachtung über die Natur, vom Herrn Karl Bonnet*, 2nd ed. (Johann Friedrich Junius, Leipzig, 1772), p. 7.

3.3 TITIUS VON WITTENBERG, JOHANN DANIEL, translation into German from French: *Betrachtung über die Natur, vom Herrn Karl Bonnet*, 4th ed. (Johann Friedrich Junius, Leipzig, 1783), p. 13.

138

REFERENCES

3.4 VIETH, GERHARD ULRICH ANTON, see "Notiz", *Monatliche Correspondenz zur Beförderung der Erd und Himmels Kunde* **7**, 78–80 (Becherischen Buchhandlung Verlag, Gotha, 1803), edited by Francis Xaver von Zach.

3.5 GROSSER, M., *The Discovery of Neptune* (Harvard University Press, Cambridge, Mass., 1962).

3.6 BENZENBERG, JOHANN FRIEDRICH, "Ueber das Entfernungsgesetz der Planeten und Monde von den Mittelpunkten ihrer Bahnen", *Annalen der Physik* **15**, 169–93 (Rengerischen Buchhandlung, Halle, 1803), ed. by Ludwig Wilhelm Gilbert.

3.7 BODE, JOHANN ELERT, *Anleitung zur Kenntniss des gestirnten Himmels*, 2nd ed. (Hamburg, 1772), p. 462.

3.8 BODE, JOHANN ELERT, *Anleitung zur Kenntniss des gestirnten Himmels*, 9th ed. (Nicolaischen Buchhandlung, Berlin and Stettin, 1823).

3.9 BODE, JOHANN ELERT, *Anleitung zur Kenntniss des gestirnten Himmels*, 3rd ed. (den Christian Friedrich Himburg, Berlin, 1777), p. 634.

3.10 BODE, JOHANN ELERT, in *Bildnisse jetztlebender Berliner Gelehrten mit ihren Selbstbiographieen* (Berlin, 1806), ed. by M. S. Lowe.

3.11 LALANDE, JOSEPH JÉRÔME LE FRANÇAIS DE, *Bibliographie Astronomique avec l'Histoire de l'Astronomie depuis 1781 jusqu'à 1802* (De l'Imprimerie de la République, Paris, 1803 [an XI]), p. 845.

3.12 BIOT, JEAN BAPTISTE, *Traité Élémentaire d'Astronomie Physique*, 3rd ed., Vol. 5 (Mallet-Bachelier, Paris, 1857), p. 331.

3.13 VON HUMBOLT, ALEXANDER, *Kosmos*, Vol. III (J. G. Cotta'scher Verlag, Stuttgart and Tübingen, 1850), p. 442.

3.14 GILLISPIE, C. C. (ed.), *Dictionary of Scientific Biography*, Vol. II (Charles Scribner's Sons, New York, 1970), pp. 220–1.

4.1 BODE, JOHANN ELERT, *Von dem neuen, zwischen Mars und Jupiter entdeckten achten Hauptplaneten des Sonnensystems* (Himburgischen Buchhandlung, Berlin, 1802), pp. 43–44. Also see p. 1, containing a letter by Piazzi.

4.2 SHAPLEY, H. and HOWARTH, H. E., *A Source Book in Astronomy* (McGraw-Hill Book Co., Inc., New York, 1929), pp. 180–2.

4.3 HERSCHEL, WILLIAM, from Herschel's manuscripts, reprinted in *The Scientific Papers of Sir William Herschel*, Vol. I (Royal Society, London, 1912), p. xxix.

4.4 LEXELL, ANDERS JEAN, "Recherches sur la nouvelle planète découverte par Herschel et nommée Georgium Sidus", *Acta Academiae Scientiarum Imperialis Petropolitanae* (Saint Peterburg), *Mémoires 1780*, **4**, 303 (1781). See p. 307.

4.5 FIXLMILLNER, PATER PLACIDUS, "Untersuchung der Elemente der Wahren Laufbahn des neuen Planeten", *Astronomische Jahrbuch (für das Jahr 1787)* **12**, 247–50 (George Jacob Decker, Königl. Hofbuchdrucker, Berlin, 1784), ed. by J. E. Bode. See p. 249.

4.6 VON ZACH, FRANCIS XAVER, "Über einen zwischen Mars und Jupiter längst vermutheten nun wahrscheinlich entdeckten neuen Hauptplaneten unseres Sonnen-Systems", *Monatliche Correspondenz zur Beförderung der Erd und Himmels Kunde* **3**, 592–606 (Becherischen Buchhandlung Verlag, Gotha, 1801), edited by the author.

REFERENCES

4.7 HEGEL, GEORG WILHELM FRIEDRICH, *Dissertatio philosophica de Orbitis Planetarum* (Pro licentia docendi, Jenae, 1801). This has been reprinted in H. Glockner, *Georg Wilhelm Friedrich Hegel Sämtliche Werke*, Vol. 1 (Fr. Frommanns Verlag, Stuttgart, 1941), pp. 3–29.

4.8 ROSENKRANZ, KARL, 1805–79, *Georg Wilhelm Friedrich Regel's Leben* (Berlin, Duncker und Humbolt, 1844), pp. 151–5.

4.9 HERSCHEL, WILLIAM, "Observations on the two lately discovered celestial bodies", *Philosophical Transactions of the Royal Society* **92**, 213–32 (London, 1802). See p. 225.

5.1 WURM, VIKARIUS (JOHANN FRIEDRICH), "Verschiedene astronomische Bemerkungen und eine Abhandlung über mögliche Planeten und Kometen unsers Sonnensystems", *Astronomisches Jahrbuch (für das Jahr 1790)* **15**, 162–73 (George Jacob Decker, Königl. Hofbuchdrucker, Berlin, 1787), ed. by J. E. Bode.

5.2 GAUSS, KARL FRIEDRICH, 1802 letter to von Zach, quoted in Francis Xaver von Zach, "Fortgesetzte Nachrichten über den neuen Haupt-Planeten unseres Sonnen-Systems, Pallas Olbersiana", *Monatliche Correspondenz zur Beförderung der Erd und Himmels Kunde* **6**, 499–505 (Becherischen Buchhandlung Verlag, Gotha, 1802), ed. by von Zach. The letter has been reprinted in *Carl Friedrich Gauss Werke*, Vol. VI (Königlichen Gesellschaft der Wissenschaften, Göttingen, 1874), pp. 230–1.

5.3 WURM, VIKARIUS (JOHANN FRIEDRICH), "Ueber die vermeinte harmonische Progression in den Planeten-Abständen, als Nachtrag zur M. C. 1802 Novbr. S. 504", *Monatliche Correspondenz zur Beförderung der Erd und Himmels Kunde* **7**, 74–78 (Becherischen Buchhandlung Verlag, Gotha, 1803), ed. by Francis Xaver von Zach.

5.4 BAILEY, J. M., "The moon may be a former planet", *Nature* **223**, 251–3 (1969).

5.5 CLAIRAUT, ALEXIS CLAUDE, "Mémoire sur la comète de 1682 addressée à MM. les auteurs de *Journal des Sçavans* par M. Clairaut", *Journal des Sçavans*, 88–96 (Paris, 1759).

5.6 GILBERT, LUDWIG WILHELM, "Ist der Ophion, (ein Planet jenseits der Uranusbahn), ein noch unbekannter Weltkörper?", *Annalen der Physik* **11**, 482–5 (Rengerischen Buchhandlung, Halle, 1802), ed. by the author.

5.7 HUSSEY, THOMAS JOHN, letter to Airy, reprinted in George Biddel Airy, "Account of some circumstances historically connected with the discovery of the planet exterior to Uranus", *Mon. Not. R. Astron. Soc.* **7**, 121–44 (1846). Same title elsewhere, including *Astronomische Nachrichten* **25**, Nos. 585 and 586, 133–48 and 149–60 (1847).

5.8 VALZ, ÉLIX BENJAMIN, letter on Halley's Comet in *Académie des Sciences, Paris, Comptes Rendus* **1**, 130–1 (1835).

5.9 NICOLAI, FRIEDRICH BERNHARD GOTTFRIED, "Schreiben des Hernn Hofraths Nicolai an den Herausgeber", *Astronomische Nachrichten* **13**, No. 294, cols. 89–95 (Altona, 1836). See col. 94. Reprinted by Johnson Reprint Corp., New York, 1964.

5.10 WARTMANN, LOUIS FRANÇOIS, "Lettre de M. Wartmann, de Genève à M. Arago, sur un astre ayant l'aspect d'une étoile et qui cependant était doué d'un mouvement propre", *Académie des Sciences, Paris, Comptes Rendus* **2**, 307–11 (1836). See p. 311.

5.11 ADAMS, JOHN COUCH, "An explanation of the observed irregularities in the motion of Uranus, on the hypothesis of disturbances caused by a more distant planet; with a determination of the mass, orbit, and position of the disturbing body", in *Nautical Almanac for the Year 1851, Appendix* (London, 1846) pp. 265–93; also with same title in *Mon. Not. R. Astron. Soc.* **7**, 149–52 (1847) and *Mem. R. Astron. Soc.* **16**, 427–60 (1847).

5.12 LEVERRIER, URBAIN JEAN JOSEPH, "Premier mémoire sur la théorie d'Uranus", *Académie des Sciences, Paris, Comptes Rendus* **21**, 1050–5 (1845); "Sur la planète qui produit les anomalies observées dans le mouvement d'Uranus. Détermination de sa masse, de son orbite et de sa position actuelle", *ibid.* **23**, 428–38, 657–62 (1846).

5.13 WALKER, SEARS COOK, "Investigations which led to the detection of the coincidence between the computed place of the Planet Leverrier and the observed place of a star recorded by Lalande in May, 1795", paper read Feb. 13th, 1847, and published in *Trans. Amer. Phil. Soc.* (N.S.) **10**, 141–53 (Philadelphia, 1853); letter dated May 3, 1847 on the elements of the planet Neptune, *Proc. Amer. Phil. Soc.* **4**, 332–5 (Philadelphia, 1847); reported in Europe by Edward Everett, "On the new planet Neptune", *Astronomische Nachrichten* **25**, No. 599, cols. 375–88 (Altona, 1847). Reprinted by Johnson Reprint Corp., New York, 1964.

5.14 LYTTLETON, R. A., "A short method for the discovery of Neptune", *Mon. Not. R. Astron. Soc.* **118**, 551–9 (1958); "The rediscovery of Neptune", in *Vistas in Astronomy* **3**, 25–46 (Pergamon Press, London, 1960), ed. by Arthur Beer; "The discovery of Neptune", in the author's *Mysteries of the Solar System* (Oxford University Press, London, 1968), chap. 7, p. 215.

5.15 TOMBAUGH, C. W., "The trans-Neptunian planet search", in chap. 2, p. 12 of ref. **(5.16)**.

5.16 KUIPER, G. P. and MIDDLEHURST, B. M. (eds.), *The Solar System*, Vol. III: *Planets and Satellites* (University of Chicago Press, Chicago, 1961).

5.17 LEVERRIER, URBAIN JEAN JOSEPH, "Lettre de M. Le Verrier à M. Faye, sur la théorie de Mercure, et sur le mouvement du périhélie de cette planète", *Académie des Sciences, Paris, Comptes Rendus* **49**, 379–83 (1859). Also see the following article on pp. 383–5: "Remarques de M. Faye à l'occasion de la lettre de M. Le Verrier".

5.18 "Romance of the new planet", no author given, in *North British Review* **33**, 1–20 (1860).

5.19 LESCARBAULT D'ORGÈRES, M. LE DOCTEUR, "Variétés. Découverte d'une nouvelle planète entre Mercure et le Soleil", *Cosmos* **16**, 22–28 (1860); "Variétés. Passage d'une planète sur le disque du soleil, observé à Orgères (Eure-et-Loir)", *ibid.* **16**, 50–56 (1860).

5.20 CHAMBERS, GEORGE FREDERICK, *Descriptive Astronomy* (Clarendon Press, Oxford, 1867), chap. 3, p. 46.

5.21 CHAMBERS, GEORGE FREDERICK, *Handbook of Descriptive and Practical Astronomy*, Vol. I, 4th ed. (Clarendon Press, Oxford, 1889), chap. 3, p. 75.

REFERENCES

5.22 HODGSON, R. G., "The Search for Vulcan", Pts. I and II, *Review of Pop. Astron.* **63**, No. 556, 10–12 (April 1969); *ibid.* **63**, No. 557, 10–11 (June 1969).

5.23 JENKINS, BENJAMIN GEORGE, "Vulcan and Bode's Law", *Nature* **19**, 74 (1879).

5.24 LEVERRIER, URBAIN JEAN JOSEPH, "Théorie nouvelle du mouvement de la planète Neptune: Remarques sur l'ensemble des théories des huit planètes principales: Mercure, Vénus, la Terre, Mars, Jupiter, Saturne, Uranus et Neptune; par M. Le Verrier", *Académie des Sciences, Paris, Comptes Rendus* **79**, 1421–7 (1874). See p. 1424.

6.1 CHALLIS, JAMES, "On the extension of Bode's empirical law of the distances of the planets from the Sun, to the distances of the satellites from their respective primaries", *Trans. Camb. Phil. Soc.* **3**, 171–83 (1830).

6.2 KIRKWOOD, DANIEL, "On the theory of meteors", *Proc. Amer. Assoc. for the Adv. of Sci.* (Buffalo, New York, 1866 meeting) **15**, 8–14 (1867).

6.3 KIRKWOOD, DANIEL, "The asteroids between Mars and Jupiter", *Smithsonian Institution, Annual Report for 1876*, pp. 358–71 (1877). This article is also reprinted in part on pp. 305–7 of ref. (4.2).

6.4 KIRKWOOD, DANIEL, letter dated July 4, 1849, *Proc. Amer. Assoc. for the Adv. of Sci.* (Cambridge, Mass., 1849 meeting) **2**, 208–11 (1850). This and the following ref. (**6.5**) are part of an entire article that goes from pages 207–21 entitled "Fifth Day, August 18, 1849. Section of Mathematics, Physics, and Astronomy".

6.5 WALKER, SEARS COOK, a reply to the letter in the preceding ref. (**6.4**), *Proc. Amer. Assoc. for the Adv. of Sci.* (Cambridge, Mass., 1849 meeting) **2**, 211 (1850); KIRKWOOD, DANIEL, another letter dated July 31, 1849, *ibid.* **2**, 211–12 (1850). The discussion ends with an examination of Kirkwood's analogy on pp. 212–21.

6.6 WALKER, S. C., "Examination of Kirkwood's analogy", *Amer. J. of Sci. and Arts* (2nd Ser.) **10**, 19–26 (1850); GOULD, Jr., BENJAMIN APTHORP (1824–96), "On Kirkwood's analogy", *ibid.* **10**, 26–31 (1850); LOOMIS, ELIAS (1811–89), "On Kirkwood's law of the rotation of the primary planets", *ibid.* **11**, 217–23 (1851); CHERRIMAN, J. BRADFORD, "On Kirkwood's analogy", *ibid.* **14**, 9–10 (1852); KIRKWOOD, DANIEL, "On certain analogies in the solar system", *ibid.* **14**, 210–19 (1852).

6.7 KIRKWOOD, DANIEL, "On certain harmonies of the solar system", *Amer. J. of Sci. and Arts* (2nd Ser.) **38**, 1–18 (1864).

6.8 ALEXANDER, STEVEN, "On some special arrangements of the solar system, which seem to confirm the nebular hypothesis", *Proc. Amer. Assoc. for the Adv. of Sci.* (Albany, 1856 meeting) **10**, pt. II, 223 (1857).

6.9 KIRKWOOD, DANIEL, "On the mean distances of the periodic comets", *Proc. Amer. Assoc. for the Adv. of Sci.* (Baltimore, 1858 meeting) **12**, 10–14 (1859).

6.10 PEIRCE, BENJAMIN, "Mathematical investigation of the fractions which occur in phyllotaxis", *Proc. Amer. Assoc. for the Adv. of Sci.* (Cambridge, Mass., 1849 meeting) **2**, 444–7 (1850).

6.11 KIRKWOOD, DANIEL, "On the formation and primitive structure of the solar system", *Proc. Amer. Phil. Soc.* **12**, 163–7 (Philadelphia, 1871). Reprinted by Kraus Reprint Corp., 1967.

6.12 ALEXANDER, STEVEN, "Statement and exposition of certain harmonies in the solar system", *Smithsonian Contributions to Knowledge* **21**, art. 2, 104 pp. (1876).

6.13 ALEXANDER, STEVEN, "On the laws of extreme distances in the solar system", *Astronomische Nachrichten* **91**, cols. 225–30 (1878).

6.14 CHASE, PLINY EARLE, "Note on planeto-taxis", *Proc. Amer. Phil. Soc.*, **13**, 143–5 (Philadelphia, 1873). Reprinted by Kraus Reprint Corp., 1967.

6.15 CHASE, PLINY EARLE, "Recent confirmation of an astronomical prediction", *Proc. Amer. Phil. Soc.* **13**, 470–7 (1873). Reprinted by Kraus Reprint Corp., 1967.

6.16 CHASE, PLINY EARLE, "The music of the spheres", *English Mechanic and World of Science* **26**, 87–88 (1877).

6.17 CHAMBERS, GEORGE FREDERICK, ref. **(5.21)**, Vol. I, p. 67.

6.18 BOHLIN, KARL, "Relationer mellan distanserna inom Saturnus-systemet", *Öfversigt af Kongl. Vetenskaps-Akademiens Förhandlingar* (Stockholm) **54**, 389–98 (1897).

6.19 CHARLIER, C. V. L., "Das Bodesche Gesetz und die sogenannten intramerkuriellen Planeten", *Astronomische Nachrichten* **193**, No. 4623, cols. 169–72 (1913).

6.20 ROCHE, ÉDOUARD, "Mémoire sur la figure d'une masse fluide, soumise à l'attraction d'un point éloigné", Pts. 1, 2, and 3 *Académie des Sciences et Lettres de Montpellier, Mémoires de la Section des Sciences* **1**, 243–62 (1849); *ibid.* **1**, 333–48 (1850); *ibid.* **2**, 21–32 (1851).

6.21 JEANS, Sir JAMES HOPWOOD, *Astronomy and Cosmogony* (Cambridge University Press, Cambridge, 1928), chaps. VIII pp. 217 ff. and IX pp. 247 ff.

6.22 FRITSCH, JOHANN HEINRICH, "Ueber das in den Abständen der Planetenbahnen sich findende Verhältniss, und über die Bestimmung der Länge aus Jupiters-mondfinsternisse", *Astronomische Jahrbuch (für das Jahr 1806)* **31**, 224–32 (1803), ed. by J. E. Bode.

6.23 VON ZACH, FRANCIS XAVER, "Ueber die relative Lage der Planetenbahnen unter sich", *Monatliche Correspondenz zur Beförderung der Erd und Himmels Kunde* **28**, 389–95 (1813), ed. by the author.

6.24 DRACH, S. M., "On some numerical relations of the solar system", *Philosophical Magazine* (3rd Ser.) **18**, 37–41 (1841).

6.25 REYNAUD, J., "Note sur les distances respectives des orbites des planètes", *Académie des Sciences, Paris, Comptes Rendus* **47**, 957–8, 1074–6 (1858).

6.26 TEBAY, SEPTIMUS, "On the law of Bode, and on the rotation of a heavenly body", *Philosophical Magazine* (4th Ser.) **15**, 206–12 (1858).

6.27 NEALE, EDWARD VANSITTART, "On the numerical relations of the distances between the planets and the Sun", *Philosophical Magazine* (4th Ser.) **26**, 462–6 (1863).

6.28 Three letters with only the authors' initials: A. —, H. E., "Distances of satellites", *English Mechanic and World of Science* **23**, 198 (1876); G. —, J. B., "Mean distances of the planets", *ibid.* **23**, 333 (1876); E. —, W. C., "Mean distances of the planets", *ibid.* **23**, 355 (1876).

6.29 GAUSSIN, LOUIS, "Lois concernant la distribution des astres du système solaire", *Académie des Sciences, Paris, Comptes Rendus* **90**, 518–20, 593–6 (1880).

6.30 Ref. **(1.12)**, Vol. 2, cols. 654–72.

6.31 Utting, James, "On a planetary analogy; or a law of motion pervading and connecting all the planetary orbits", *Philosophical Magazine and Journal* **62**, 119–21, 214 (postscript) (1823).

6.32 Billy, "Sur une analogie planétaire, ..." and "Supplément au mémoire de M. Utting, ...", reports on ref. **(6.31)** in *Bulletin des Sciences Mathématiques, Astronomiques, Physiques et Chimiques, Paris* **1**, 203–4, 204–5 (1824); "Note additionnelle...", *ibid.* **1**, 279–80 (1824), reports on two refutations of ref. **(6.31)** one of which is contained in ref. **(6.33)**.

6.33 Prévost, Pierre (1751–1839), "Note sur une analogie remarquée dans le système planétaire", *Bibliothèque Universelle des Sciences Belles Lettres et Arts, Genève* **24**, 252–3 (1823); "Note *additionnelle* sur une analogie remarquée dans le système planétaire", *ibid.* **25**, 30 (1824).

6.34 Gergonne, Joseph Diez (1771–1859), "Sur une loi prétendue nouvelle des mouvements célestes", *Annales de Mathématiques Pures et Appliquées, Paris* **14**, 272–5 (1824).

7.1 Blagg, Mary Adela, "On a suggested substitute for Bode's Law", *Mon. Not. R. Astron. Soc.* **73**, 414–22 (1913).

7.2 Allen, C. W., *Astrophysical Quantities* (Athlone Press, London, 1955), chap. 8.

7.3 Richardson, D. E., "Distances of planets from the Sun and of satellites from their primaries in the satellite systems of Jupiter, Saturn, and Uranus", *Pop. Astron.* **53**, 14–26 (1945).

7.4 Roy, A. E., "Miss Blagg's formula", *J. Brit. Astron. Assoc.* **63**, 212–15 (1953).

7.5 Schuette, C. H., "Two new families of comets", *Pop. Astron.* **57**, 176–82 (1949); "Drei weitere Mitglieder der Transplutokometenfamilie", *Acta Astronomica* **15**, 11–13 (1965).

7.6 Bowell, E. L. G. and Wilson, L., "Arguments for the presence of undiscovered satellites", *Nature* **216**, 669 (1967).

7.7 Bigg, E. K., "New satellites for Jupiter", *Astrophysical Letters* **3**, 77–79 (1969).

7.8 Richardson, D. E., "Distribution of planet mass in the solar system", *Pop. Astron.* **52**, 497–505 (1944).

8.1 Laplace, Pierre Simon Marquis de, *Traité de Méchanique Céleste*, Vol. I, bk. 2, chap. VII, §65 and Vol. IV, bk. 8, chap. II, §4 (Chez J. B. M. Duprat, Paris, 1798 [an. VII] and 1805 [an. XIII]). Translated into English as *Méchanique Céleste* by Nathaniel Bowditch (1773–1838), with comment (Hilliard, Gray, Little, and Wilkins, Pub., Boston, 1829 and 1839); reprinted as *Celestial Mechanics* (Chelsea Publ. Co., New York, 1966). Also see ref. **(8.2)**.

8.2 Tisserand, François Félix, *Traité de Méchanique Céleste*, Vol. IV (Gauthier-Villars et Fils, Paris, 1896), chap. II, §10, p. 25.

8.3 Roy, A. E. and Ovenden, M. W., "On the occurrence of commensurable mean motions in the solar system", *Mon. Not. R. Astron. Soc.* **114**, 232–41 (1954).

8.4 Goldreich, P., "An explanation of the frequent occurrence of commensurable mean motions in the solar system", *Mon. Not. R. Astron. Soc.* **130**, 159–81 (1965).

8.5 DERMOTT, S. F., "On the origin of commensurabilities in the solar system—II. The orbital period relation", *Mon. Not. R. Astr. Soc.* **141**, 363–76 (1968).

8.6 ALFVÉN, H., "On the origin of the solar system", *Quart. J. R. Astron. Soc.* **8**, 215–26 (1967).

8.7 TER HAAR, D. and CAMERON, A. G. W., "Historical review of theories of the origin of the solar system", in *Origin of the Solar System* (Academic Press, New York, 1963), ed. by R. Jastrow and A. G. W. Cameron, pp. 1–37.

8.8 MOLCHANOV, A. M., "The resonant structure of the solar system. The law of planetary distances", *Icarus* **8**, 203–15 (1968).

8.9 BACKUS, G. E., "Critique of 'The resonant structure of the solar system' by A. M. Molchanov", *Icarus* **11**, 88–92 (1969).

8.10 HENON, M., "A comment on 'The resonant structure of the solar system', by A. M. Molchanov", *Icarus* **11**, 93–94 (1969).

8.11 DERMOTT, S. F., "On the origin of commensurabilities in the solar system—III. The resonant structure of the solar system", *Mon. Not. R. Astron. Soc.* **142**, 143–9 (1969).

8.12 MOLCHANOV, A. M., "Resonances in complex systems: a reply to critiques", *Icarus* **11**, 95–103 (1969); "The reality of resonances in the solar system", *ibid.* **11**, 104–13 (1969).

8.13 NIETO, M. M., "Conclusions about the Titius–Bode Law of Planetary Distances", *Astron. and Astrophys.* **8**, 105–11 (1970).

9.1 WILLIAMS, I. P. and CREMIN, A. W., "A survey of theories relating to the origin of the solar system", *Quart. J. R. Astron. Soc.* **9**, 40–62 (1968).

9.2 WOOLFSON, M. M., "The evolution of the solar system", *Rep. Prog. Phys.* **32**, 135–85 (1969).

9.3 HOYLE, F., *Frontiers of Astronomy* (William Heinemann, Ltd., London, 1955), chap. 6.

9.4 HOYLE, F., "On the origin of the solar nebula", *Quart. J. R. Astron. Soc.* **1**, 28–55 (1960).

9.5 ALFVÉN, H., "Remarks on the rotation of a magnetized sphere, with application to solar rotation", *Arkiv för Mat., Astr. och Fysik* **28A**, No. 6 (1942).

9.6 LÜST, R. and SCHLÜTER, A., "Drehimpulstransport durch Magnetfelder und die Abbremsung rotierender Sterne", *Zeit. für Astroph.* **38**, 190–211 (1955).

9.7 CHANDRASEKHAR, S., *An Introduction to the Study of Stellar Structure* (University of Chicago Press, Chicago, 1939), chap. XII.

10.1 POISSON, SIMÉON DENIS, "Mémoire sur les inégalités séculaires des moyens mouvements des planètes", *J. École Polytechnique* **15**, 1–56 (1809).

10.2 LAGRANGE, JOSEPH LOUIS COMTE DE, "Sur l'altération des moyens mouvements des planètes", *Königliche Akademie der Wissenschaften, Berlin, Nouveaux Mémoires 1776 (Nouveaux Mémoires de l'Académie Royale des Sciences et Belles Lettres. Année 1776)*, 199–213 (Georges Jacques Decker, 1779). Reprinted in *Œuvres de Lagrange*, Vol. IV (Gauthier-Villars, Paris, 1869), pp. 255–71.

REFERENCES

10.3 Tisserand, François Félix, "Mémoire sur un point important de la théorie des perturbations planétaires", *Mémoires de l'Académie des Sciences, Inscriptions et Belles-Lettres de Toulouse* (Ser. 7) **7**, 374–88 (1875).

10.4 Hagihara, Y., "A proof of Poisson's theorem on the invariability of the major axis of planetary orbits", *Japan. J. Astron. Geophys.* **21**, 9–27 (1944); "The stability of the solar system", in chap. 4, p. 95 of ref. **(5.16)**. See §2.

10.5 Hills, J. G., "Dynamic relaxation of planetary systems and Bode's Law", *Nature* **225**, 840–2 (1970).

10.6 Darwin, Sir George Howard, "On the tidal friction of a planet attended by several satellites, and on the evolution of the solar system", *Phil. Trans. Roy. Soc.* **172**, 491–535 (1881). See §9. This article was reprinted in *Scientific Papers by Sir George Howard Darwin*, Vol. II (Cambridge University Press, Cambridge, 1908), pp. 406–58.

10.7 Roy, A. E. and Ovenden, M. W., "On the occurrence of commensurable mean motions in the solar system. II. The Mirror Theorem", *Mon. Not. R. Astron. Soc.* **115**, 296–309 (1955).

10.8 Brouwer, Dirk (1902–66) and Clemence, G. M., "Orbits and masses of planets and satellites", in chap. 3, p. 31 of ref. **(5.16)**. See Sec. 3.4. Also see *Celestial Mechanics* (Academic Press, New York, 1961), p. 529.

10.9 Dermott, S. F., "On the origin of commensurabilities in the solar system—I. The tidal hypothesis", *Mon. Not. R. Astron. Soc.* **141**, 349–61 (1968).

10.10 Jeffreys, H., "The secular accelerations of satellites", *Mon. Not. R. Astron. Soc.* **117**, 585–9 (1957).

10.11 Hagihara, Y., "The stability of the solar system", in chap. 4, p. 95 of ref. **(5.16)**. See §5, p. 124.

10.12 Moser, J., "Stabilitätsverhalten kanonischer Differentialgleichungssysteme", *Nachrichten Akad. Wiss. Göttingen, IIa Math.-Phys.-Chem. Ab.*, Nr. 6, 87–120 (1955); "Stability of the asteroids", *Astron. J.* **63**, 439–43 (1958).

10.13 Jefferys, W. H., "Nongravitational forces and resonances in the solar system", *Astron. J.* **72**, 872–5 (1967).

10.14 Committee on Celestial Mechanics, "Celestial mechanics", in *Transactions of the International Astronomical Union*, **14A**, *Reports on Astronomy* (D. Reidel Publ. Co. Dordrecht, Holland, 1970), ed. by C. de Jager, chap. 7, p. 19.

12.1 Birkeland, Olaf Kristian, "Sur l'origine des planètes et de leurs satellites", *Académie des Sciences, Paris, Comptes Rendus* **155**, 892–5 (1912).

12.2 Berlage, Jr., H P., "Versuch einer Entwicklungsgehscichte der Planeten", Ergänzungsheft.(Supplement) to *Gerlands Beiträge zur Geophysik* **17**(1927).

12.3 Berlage, Jr., H. P., "On the electrostatic field of the Sun, due to its corpuscular rays", *Proc. Koninkl. Ned. Akad. Wetenschap. (Amsterdam)* **33**, 614–18 (1930); "On the electrostatic field of the Sun as a factor in the evolution of the planets", *ibid.* **33**, 719–22 (1930).

12.4 Berlage, Jr., H. P., "A study of the systems of satellites from the standpoint of the disk-theory of the origin of the planetary system", *Annalen v.d. Bosscha-Sterrenwacht (Lembang-Java)* **4**, Miscell. papers No. 7, 79–94 (1934).

12.5 Berlage, Jr., H. P., "Spontaneous development of a gaseous disc revolving round the Sun into rings and planets", Pts. I and II, *Proc. Koninkl. Ned. Akad. Wetenschap. (Amsterdam)* **43**, 532–41, 557–66 (1940).

12.6 ALFVÉN, H., "On the cosmogony of the solar system", Pts. I, II, and III, *Stockholms Obs. Ann.* **14**, No. 2 (1942); *ibid.* **14**, No. 5 (1943); *ibid.* **14**, No. 9 (1946).

12.7 ALFVÉN, H., *On the Origin of the Solar System* (Clarendon Press, Oxford, 1954).

12.8 ALFVÉN, H., "Cosmical electrodynamics", *Amer. J. Phys.* **28**, 613–18 (1960). Also see "On the origin of the solar system", in the *New Scientist* **7**, 1188–91 (May 12, 1960).

12.9 ALFVÉN, H. and WILCOX, J. M., "On the origin of the satellites and the planets", *Astrophys. J.* **136**, 1016–22 (1962).

12.10 ALFVÉN, H., "On the mass distribution in the solar system", *Astrophys. J.* **136**, 1005–15 (1962); "On the early history of the Sun and the formation of the solar system", *ibid.* **137**, 981–90 (1963).

13.1 SCHMIDT, O. J., "On the origin of visual binary stars and the character of their orbits", *Comptes Rendus (Doklady) Acad. Sci. URSS* **44**, 8–12 (1944); "A meteoric theory of the origin of the Earth and planets", *ibid.* **45**, 229–33 (1944).

13.2 SCHMIDT, O. J., *A Theory of the Origin of the Earth; Four Lectures* (Lawrence & Wishart, London, 1959).

13.3 SCHMIDT, O. J., "On the law of planetary distances", *Comptes Rendus (Doklady) Acad. Sci. URSS* **52**, 667–72 (1946).

13.4 RANDIĈ, L., "Schmidt's theory of the origin of visual binary stars and of the solar system", *Observatory* **70**, 217–22 (1950).

13.5 ROBERTSON, H. P., "Dynamical effects of radiation in the solar system", *Mon. Not. R. Astron. Soc.* **97**, 423–38 (1937).

13.6 EGYED, L., "Dirac's cosmology and the origin of the solar system", *Nature* **186**, 621–2 (1960).

13.7 DIRAC, P. A. M., "The cosmological constants", *Nature* **139**, 323 (1937); "A new basis for cosmology", *Proc. Roy. Soc. (London)* **A165**, 199–208 (1938).

13.8 SANDAGE, A., "The distance scale", in *Problems of Extra-Galactic Research, IAU Symposium* No. 15, Aug. 10–12, 1961 (Macmillan Co., New York, 1962), ed. by G. C. McVittie, pp. 359–78.

13.9 HOLMBERG, E., "A study of external galaxies", *Arkiv för Astronomi* **3**, 387–438 (1964). Also printed as *Medd. Uppsala Astr. Obs.*, No. 148.

13.10 TELLER, E., "On the change of physical constants", *Phys. Rev.* **73**, 801–2 (1948).

13.11 GAMOW, GEORGE (1904–68), "Electricity, gravity, and cosmology", *Phys. Rev. Letters* **19**, 759–61 (1967).

13.12 SCHOPF, I. W. and BARGHOORN, E. S., "Alga-like fossils from the early Precambrian of South Africa", *Science* **156**, 508–12 (1967).

13.13 POCHODA, P. and SCHWARZSCHILD, M., "Variation of the gravitational constant and the evolution of the Sun", *Astrophys. J.* **139**, 587–93 (1964).

13.14 GAMOW, GEORGE, "Does gravity change with time?", *Proc. Natl. Acad. Sci. (USA)* **57**, 187–93 (1967).

13.15 WILKINSON, D. H., "Do the 'Constants of Nature' change with time?", *Phil. Mag.* (8th Ser.) **3**, 582–5 (1958).

13.16 DYSON, F. J., "Time variation of the charge of the proton", *Phys. Rev. Letters* **19**, 1291–3 (1967); PERES, A., "Constancy of the fundamental electric charge", *ibid.* **19**, 1293–4 (1967); BACHALL, J. N. and SCHMIDT, M., "Does the fine-structure constant vary with cosmic time?", *ibid.* **19**, 1294–5 (1967).

13.17 WOOLFSON, M. M., "Origin of the solar system", *Nature* **187**, 47–48 (1960).

13.18 WOOLFSON, M. M., "A capture theory of the origin of the solar system", *Proc. Roy. Soc. (London)* **A282**, 485–507 (1964).

13.19 PENDRED, B. W. and WILLIAMS, I. P., "The formation of the planets", *Icarus* **8**, 129–37 (1968).

13.20 PENDRED, B. W. and WILLIAMS, I. P., "Planetary masses and distances", *Astrophysics and Space Science* **5**, 420–4 (1969).

13.21 GOLDREICH, P. and PEALE, S., "Spin–orbit coupling in the solar system", *Astron. J.* **71**, 425–38 (1966); "Spin–orbit coupling in the solar system. II. The resonant rotation of Venus", *ibid.* **72**, 662–8 (1967); "The obliquity of Venus", *ibid.* **75**, 273–84 (1970).

13.22 ANTAL, J. A., Correspondence on "Bode's Law", *Nature* **227**, 642 (1970).

13.23 ROOKES, D., Correspondence on "Bode's Law", *Nature* **227**, 981 (1970).

13.24 DOLE, S. H., "Computer simulation of the formation of planetary systems", *Icarus* **13**, 494–508 (1970).

13.25 JEANS, Sir JAMES HOPWOOD, *The Dynamical Theory of Gases*, 2nd ed. (Cambridge University Press, Cambridge, 1916), chap. xv, pp. 348 ff.

14.1 BERLAGE, Jr., H. P., "On the structure and internal motion of the gaseous disc constituting the original state of the planetary system", *Proc. Koninkl. Ned. Akad. Wetenschap. (Amsterdam)* **35**, 553–62 (1932).

14.2 BERLAGE, Jr., H. P., "Viscosity and steady states of the disc constituting the embryo of the planetary system", *ibid.* **37**, 221–32 (1934).

14.3 BERLAGE, Jr., H. P., "The theorem of minimum loss of energy due to viscosity in steady motion and the origin of the planetary system from a rotating gaseous disc", *ibid.* **38**, 857–62 (1935).

14.4 BERLAGE, Jr., H. P., "The disc theory of the origin of the solar system", *ibid.* **51**, 796–806 (1948).

14.5 BERLAGE, Jr., H. P., "Types of satellite systems and the disc theory of the origin of the planetary system", *ibid.* **51**, 965–8 (1948); "Some remarks on the internal constitution of the bodies of the solar system", *ibid.* **B54**, 344–9 (1951); "On the composition of the bodies of the solar system", *ibid.* **B56**, 45–55 (1953).

14.6 BERLAGE, Jr., H. P., "The masses of planets and satellites derived from the disk theory of the origin of the solar system", *ibid.* **B56**, 56–66 (1953); "Remarks on the origin of satellites in general and on the metamorphosis of the systems of Neptune and the Earth in particular", *ibid.* **B57**, 452–63 (1954); "The basic scheme of any planetary or satellite system", *ibid.* **B60**,

75–87 (1957); " The basic scheme of any planetary or satellite system corrected and reanalyzed", *ibid.* **B62,** 63–83 (1959); "On accretional instability, the state, leading to the transformation of a gaseous disk, rotating in quasi-steady motion round a massive centre, into a set of concentric rings of particulate matter", Pts. I and II, *ibid.* **B65,** 199–210, 211–20 (1962); "Remarks on the position of our Moon and of Saturn's tenth satellite in the basic scheme of satellite systems", *ibid.* **B70,** 363–6 (1967).

14.7 BERLAGE, Jr., H. P., "The origin of the Moon II", in *Mantles of the Earth and Terrestrial Planets* (Interscience Publishers, London, 1967), ed. by S. K. Runcorn, pp. 241–50.

14.8 VON WEIZSÄCKER, C. F., "Über die Entstehung des Planetensystems", *Zeit. für Astroph.* **22,** 319–55 (1943); "Zur Kosmogonie", *ibid.* **24,** 181–206 (1947).

14.9 GAMOW, GEORGE and HYNEK, J. A., "A new theory by C. F. von Weizsäcker of the origin of the planetary system", *Astrophys. J.* **101,** 249–54 (1945).

14.10 CHANDRASEKHAR, S., "On a new theory of Weizsäcker on the origin of the solar system", *Rev. Mod. Phys.* **18,** 94–102 (1946).

14.11 CAMERON, A. G. W., "Cosmic abundances of the elements", Yeshiva University unpublished report (1967).

14.12 STRÖMGREN, B., "On the helium and hydrogen content of the interior of the stars", *Astrophys. J.* **87,** 520–34 (1938).

14.13 BIERMANN, L., "Über die chemische Zusammensetzung der Sonne", *Zeit. für Astroph.* **22,** 244–64 (1943).

14.14 TER HAAR, D., "Studies on the origin of the solar system", *D. Kgl. Danske Vidensk. Selskab, Mat.-fys. Medd.* **25,** No. 3 (1948). Also under the same title as University of Leiden thesis (1948).

14.15 CHANDRASEKHAR, S., *Principles of Stellar Dynamics* (University of Chicago Press, Chicago, 1942), p. 156.

14.16 BATCHELOR, G. K., *The Theory of Homogeneous Turbulence* (Cambridge University Press, Cambridge, 1959).

14.17 TUOMINEN, J., "Weizsäcker's theory of the origin of the solar system and the theory of turbulence", *Annales d'Astrophysique* **10,** 179–80 (1947).

14.18 PRANDTL, LUDWIG, *Abriss der Strömungslehre* (F. Vieweg und Sohn, Braunschweig, 1931), p. 93.

14.19 VON KÁRMÁN, THEODORE, "Mechanische Ähnlichkeit und Turbulenz", *Nachrichten Akad. Wiss. Göttingen, Math.-Phys. Kl.,* 58–76 (1930).

14.20 REYNOLDS, OSBORNE, "An experimental investigation of the circumstances which determine whether the motion of water shall be direct or sinuous, and of the law of resistance in parallel channels", *Phil. Trans. Roy. Soc.* **174,** 935–82 (1883). Reprinted in the author's *Papers on Mechanical and Physical Subjects,* Vol. II (Cambridge University Press, Cambridge, 1901), paper 44, p. 51.

14.21 LANDAU, LEV DAVIDOVICH (1908–68) and LIFSHITZ, E. M., *Fluid Mechanics* (Pergamon Press, Oxford, 1959), chap. 3.

14.22 TER HAAR, D., "Further studies on the origin of the solar system", *Astrophys. J.* **111,** 179–87 (1950).

14.23 HEISENBERG, W., "Zur statistischen Theorie der Turbulenz", *Zeit. für Phys.* **124,** 628–57 (1948).

14.24 HEISENBERG, W., "On the theory of statistical and isotropic turbulence", *Proc. Roy. Soc.* **A195**, 402–6 (1948).

14.25 CHANDRASEKHAR, S., "On Heisenberg's elementary theory of turbulence", *Proc. Roy. Soc.* **A200**, 20–33 (1949).

14.26 CHANDRASEKHAR, S. and TER HAAR, D., "The scale of turbulence in a differentially rotating gaseous medium", *Astrophys. J.* **111**, 187–90 (1950).

14.27 KOLMOGOROFF, A. N., "The local structure of turbulence in an incompressible viscous fluid for very large Reynolds numbers", *Comptes Rendus (Doklady) Acad. Sci. URSS* **30**, 301–5 (1941); "On degeneration of isotropic turbulence in an incompressible viscous liquid", *ibid.* **31**, 538–40 (1941); "Dissipation of energy in the locally isotropic turbulence", *ibid.* **32**, 16–21 (1941).

14.28 ONSAGER, L., "The distribution of energy in turbulence", *Phys. Rev.* **68**, 286 (1945).

14.29 VON WEIZSÄCKER, C. F., "Das Spektrum der Turbulenz bei grossen Reynoldsschen Zahlen", *Zeit. für Phys.* **124**, 614–27 (1948).

14.30 KUIPER, G. P., "The law of planetary and satellite distances", *Astrophys. J.* **109**, 308–13 (1949); (Errata) **109**, 555 (1949).

14.31 KUIPER, G. P., "On the origin of the solar system", *Proc. Natl. Acad. Sci. (U.S.A.)* **37**, 1–14 (1951); (Errata) **37**, 233 (1951).

14.32 KUIPER, G. P., "On the origin of the solar system", in *Astrophysics, a Topical Conference* (McGraw-Hill Book Co., Inc., New York, 1951), ed. by J. A. Hynek, chap. 8, p. 357.

14.33 Ref. **(6.21)**, chap. XIII, §313, pp. 337 ff.

14.34 HATTORI, T., NAKANO, T., and HAYASHI, C., "Thermal and dynamical evolution of gas clouds of various masses", *Prog. Theor. Phys.* **42**, 781–98 (1969).

14.35 NAKANO, T., "Origin of the solar system", *Prog. Theor. Phys.* **44**, 77–98 (1970).

15.1 PALMÉN, E. and NEWTON, C. W., *Atmospheric Circulation Systems, Their Structure and Physical Interpretation* (Academic Press, New York, 1969). See chaps. 1, 2, and 17.

Author–Name Index

When the full name and years of birth and death of deceased persons were known to us, we have included them. The **boldface** type describes reference numbers, and the lightface type refers to text page numbers where the individual and/or his work is mentioned. Generally only reference numbers (and first page listings for books) are given for editors or translators of authored works. We have not listed the time that very commonly used names in the text were mentioned in terms of effects (i.e. Titius–Bode Law, Blagg–Richardson formulation, Kepler orbits, etc.), unless they were especially important. Consult the Subject Index.

Subject Index

OTHER TITLES IN THE SERIES
IN NATURAL PHILOSOPHY